NON-LINEAR PHENOMENA IN SCIENCE AND ENGINEERING

ELLIS HORWOOD SERIES IN PHYSICS AND ITS APPLICATIONS

Series Editors: JOHN W. MASON, Scientific and Technical Consultant; MALCOLM COOPER, Department of Physics, University of Warwick; and E. H. GRANT, Department of Physics, King's College, University of London

Choudhury, M. H.	**Electromagnetism**
Delaney, C.F.G.	**Electronics for the Physicist with Applications**
Dyson, N.A.	**An Introduction to Nuclear Physics, with Applications in Medicine and Biology**
Elwell, D.	**Physics for Engineers and Scientists**
Gough, W., Richards, J.P.G. and Williams, R.P.	**Vibrations and Waves**
Granier, R. and Gambini, D.-J.	**Applied Radiation Biology and Protection**
Hasnain, S.S. (editor)	**Synchrotron Radiation and Biophysics**
Hill, C.R. (editor)	**Physical Principles of Medical Ultrasonics**
Rosser, W.G.V.	**An Introduction to Statistical Physics**
Rowlands, G.	**Non-Linear Phenomena in Science and Engineering**
Scott, V.D. and Love, G. (editors)	**Quantitative Electron Probe Microanalysis**
Steward, E.G.	**Fourier Optics: An Introduction,** *2nd edition*
Whorlow, R.W.	**Rheological Techniques**

NON-LINEAR PHENOMENA IN SCIENCE AND ENGINEERING

G. ROWLANDS B.Sc., Ph.D.
Department of Physics, University of Warwick

ELLIS HORWOOD
NEW YORK   LONDON   TORONTO   SYDNEY   TOKYO   SINGAPORE

First published in 1990
Reprinted and issued in paperback for the first time 1993 by
ELLIS HORWOOD LIMITED
Market Cross House, Cooper Street,
Chichester, West Sussex, PO19 1EB, England

A division of
Simon & Schuster International Group
A Paramount Communications Company

Typeset in Times by Ellis Horwood Limited
Printed and bound in Great Britain
by Hartnolls, Bodmin, Cornwall

British Library Cataloguing in Publication Data

Rowlands, G.
Non-linear phenomena in science and engineering.
1. Nonlinear equations
I. Title
515.252
ISBN 0–13–624487–4 Hbk
ISBN 0–13–104043–X Pbk

Library of Congress Cataloging-in-Publication Data

Rowlands, George.
Non-linear phenomena in science and engineering / G. Rowlands.
p. cm. — (Ellis Horwood series in physics and its applications.)
ISBN 0–13–624487–4

1. Nonlinear theories. 2. Mathematical physics. I. Title. II. Series.
QC20.7.N6R68   1990
530.1–dc20

90–4857
CIP

Table of contents

To Inga
who bullied and cajoled me and the manuscript
to bring the book to fruition

Preface

Although Poincaré at the very beginning of the present century understood and wrote about much of what is now causing excitement in the field of non-linear science, his work was ignored by most of the scientific community until fairly recently.

Linear theories were so successful, and also so much easier to analyse, that there was great resistance to change. Not until a few phenomena which are essentially non-linear, like the chaotic solutions of the logistic map as applied to ecological problems by May or the strange attractor as found by Lorenz in his analysis of weather prediction, yielded to theoretical analysis, did the study of intrinsic non-linear phenomena change from a cottage to a major industry.

Turbulence, a truly non-linear phenomenon, had, despite heroic efforts, remained as baffling as in Poincaré's time. However, an interpretation in terms of chaos and strange attractors has given new insights into this universal effect.

One major reason for the recent advances in the understanding of non-linear phenomena has undoubtedly been the accessibility of computers and in particular the advent of good computer graphics. This has meant that the unique analytic intuition of Poincaré can be shared by those who have even a modest computer. For example the early decisive work of Lorenz in the sixties is easily repeated and in fact readily extended using a small desktop PC.

Even so, the approach taken in this book is to explore analytically simple models which preserve the essential non-linear feature rather than give a catalogue of numerical results. This is partly prejudice on the part of the author, but it is a fact that a full understanding of any phenomena can only be obtained by a careful dovetailing of analytic and numerical methods, although the mixture of the methods is a matter of personal choice.

Many of the phenomena discussed in Chapters 7 and 8 are readily and profitably explored using numerical methods. A software package written by the author and Professor J. C. Sprott is available and illustrates most of the phenomena discussed in these chapters. For details see the section on further reading at the end of the book.

A particular aspect of non-linear science which has received widespread coverage is **chaos**, which is discussed in Chapter 7, but this is just one, albeit important, aspect of non-linear science. In fact, in total contrast to the pseudo-random behaviour associated with strange attractors and chaos , non-linearity can lead to the formation of coherent structures. The best-known example is the **soliton**, which is a pulse-like disturbance which propagates in a dispersive medium without change of shape; that is, it does not disperse. Solitons are the subject of Chapter 6.

The aim of this book is to introduce some of the essential features of non-linearity and the analytic methods that can be used to study such phenomena. The subject matter dealt with has had to be selective and the selection in part reflects the author's interests. These have accrued during a research career mainly devoted to the understanding of plasma physics problems arising both in the laboratory and in nature.

Some twenty years ago I introduced an optional course on non-linear physics to third year undergraduates at the University of Warwick. Much of this was based on my research interests in plasma physics. However, in 1978 I read and was greatly impressed by Haken's book entitled *Synergetics* and it dawned on me that the material I was familiar with in plasma physics had a far more universal application. I hastily changed the course title to 'Synergetics' and have taught synergetics ever since. In the academic year of 1988/89 I had the great fortune to visit the Physics Department at the University of Wisconsin at Madison and teach a graduate course for a full semester. The course was an introduction to non-linear physics and the material presented in that course forms the basis of the present book.

Based on my experience the book should be suitable for final year undergraduates in physics and engineering in the UK and senior year and first year graduates in the USA. Mathematicians could benefit from seeing the treatment of realistic problems, although little attempt is made here to obtain rigorous results. The main mathematical requirement is a working knowledge of the calculus and hence this book should be accessible to mathematically aware chemists, biologists and economists. Although some of the illustrations are from physics, no detailed knowledge of this subject is necessary to understand the new methods being used in the study of non-linear phenomena in an ever-widening field of application.

A number of questions are given at the end of each chapter (except Chapter 1) and detailed answers given at the end of the book. These questions and answers should be seen as an integral part of the book. They were designed not only to help the understanding of the text but to expand on the more important concepts introduced in the text.

At the end of the book is a list of works and articles which have been chosen to guide the reader in further study. Again my prejudice shows in recommending my own book written with my colleague Prof. E. Infeld entitled *Non-linear waves, solitons and chaos*.

My thanks go to all the students who have taken my courses and who, by their questioning, honed my presentation, particularly Christoph Litwin; to Stewart Praegar, who not only read the manuscript and made most valuable comments but who also gave encouragement when it was needed; and finally to Hayley Gilder for typing the manuscript in the professional manner to which I have become accustomed.

1

Introduction

Until fairly recently most of the advances in physics and other sciences have been made by studying linear models, although the world around us is inherently non-linear. This has been for a good reason: linear models are both conceptually and mathematically easier to understand. For example, consider the simple pendulum. The angle of displacement, θ satisfies the equation

$$\frac{d^2\theta}{dt^2} + \omega^2 \sin\theta = 0 \ , \tag{1.1}$$

where ω^2 is the ratio of the gravitational acceleration to the length of the pendulum. Because the dependent variable θ enters this equation in a form other than being directly proportional to θ or its derivatives, it is a non-linear equation. However, if we restrict attention to small values of $\theta (\theta \leqslant 10°)$, then $\sin\theta \simeq \theta$ and the above equation reduces to

$$\frac{d^2\theta}{dt^2} + \omega^2\theta = 0 \ . \tag{1.2}$$

This is a linear equation which is readily solved to give

$$\theta = A\cos(\omega t + \chi) \ ,$$

where A and χ are constants determined by initial conditions.

The important aspect of this solution is that θ is a periodic function of t, period $2\pi/\omega$. The periodicity of a simple pendulum is said to have been first appreciated by Galileo who timed the motion of a chandelier using his pulse. Slightly more sophisticated theory shows that even for the non-linear problem, as described by

(1.1), the motion is still periodic. The period, unlike the linear case, is now a function of the amplitude, a typical non-linear phenomenon. The effect is too small for Galileo to have detected it using his pulse.

Importantly, however, the linear theory captures the *essentials* of the full problem. This process of linearization is the basis for many of the theoretical models used in science. A good example is the free electron theory of metals and semiconductors. The electrons are not, in reality, free: they interact among themselves and with the lattice. All these complicated effects, however, can be accounted for by ascribing to the electron an 'effective' mass and then treating it as free. All the physics of the interactions has been subsumed into this mass. A complicated non-linear problem has been linearized. Interestingly this procedure was only justified long after its value had been thoroughly demonstrated. It has its range of validity, and, in fact, breaks down completely in superconducting material. For such material the electron–lattice interaction is so strong that it introduces an electron–electron interaction which is attractive and greater than the Coulomb repulsion. This leads to the formation of Cooper pairs. Then instead of considering an ensemble of strongly interacting electrons and a lattice (a non-linear problem), one may consider an ensemble of weakly interacting Cooper pairs (a linear problem).

A further advantage of linear equations is their universality: many distinct models reduce to the same linear one. For example, consider the motion of a particle moving in a one-dimensional potential $V(x)$, so that the equation of motion, for a mass m, is

$$m \frac{d^2 x}{dt^2} = -\frac{\partial V}{\partial x} .$$

Suppose the potential has a minimum at $x = x_0$, then for small displacements about this point ($x = x_0 + \delta x$) one can write

$$m \frac{d^2 \delta x}{dt^2} = -V''(x_0) \, \delta x + O(\delta x^2) ,$$

where $V''(x_0)$ is the second derivative of V evaluated at $x = x_0$. (Here we have introduced the nomenclature $O(\delta x^2)$ which stands for terms of order δx^2 and all higher powers.) If we now linearize, that is, neglect $O(\delta x^2)$ terms, the above reduces to the form of (1.2) with a frequency given by $\omega^2 = V''(x_0)/m$.

It has been assumed in the past, implicitly or explicitly, that this universality would be lost if non-linear terms were retained. This gives a good reason for trying to keep linear models for as long as possible, but, as we shall see, there is a universality for non-linear models, albeit of a somewhat different nature.

A major mathematical reason for retaining linear models is the relative ease with which boundary or initial conditions can be satisfied. For example, (1.2) has two independent solutions, $A\cos\omega t$ and $B\sin\omega t$. Because the equation is linear the complete solution is simply their sum, $\theta = A\cos\omega t + B\sin\omega t$. This is complete since a second-order differential equation must have a solution which involves two, and only two, arbitrary constants. Now suppose we impose the initial conditions that $\theta = 0$ and $d\theta/dt = 1$ at $t = 0$. This leads to two algebraic equations $A = 1$ and $B = 0$ which in turn lead to a unique solution. This is not the case for non-linear equations. Even suppose one knew two solutions, θ_1 and θ_2 say, of (1.1), then $\theta = A\theta_1 + B\theta_2$ is *not* a solution of this equation, a fact very easily demonstrated by direct substitution. Furthermore, even if one knew the complete solution of the equation involving two arbitrary constants, the implementation of initial conditions leads to functional equations, not algebraic ones, to be solved for these constants.

Fortunately, in many experiments in physics, it is possible to ensure that the system remains in the linear regime. For pendulum experiments the initial amplitude is kept less than $10°$, in which case all properties can be explained in terms of the linear model (1.2). Maxwell's equations for the propagation of light in a vacuum are linear, but in other media may be non-linear. The non-linearity arises from the fact that the dielectric constant is in general a function of the light intensity. Usually the complications of non-linearity can be avoided by keeping the light intensity low. This is all to the good, *unless*, by imposing these restrictions, one misses important phenomena: not a quantitative change in response, (these can usually be accounted for by a change of parameters), but qualitative ones.

The advent of the laser made it necessary to consider non-linear effects in areas where previously linear models were sufficient. For example, a monochromatic wave, of high intensity, incident on a dielectric, produces a wave at double the frequency of the incident wave a truly non-linear phenomenon.

Although in many branches of physics non-linear phenomena could be avoided or cunningly by-passed, this is not so in engineering. It is not always possible to design structures such that only small amplitude oscillations occur. Witness the dramatic, highly non-linear behaviour which led to the Jacoma Narrows bridge disaster! Despite such examples, Western engineers preferred the 'linear fix', whilst Soviet ones, for a number of reasons, found that they had to study and treat non-linear effects much more seriously. In the late 1920s, Russian electrical engineers had encountered what is now called chaos, and appreciated that it differed in a fundamental manner from random effects associated with thermal noise. Thus in the 1950s and 1960s it was the Russian mathematicians and engineers who led the way in the study of non-linear equations — a situation to which Minorsky drew attention. His book *Non-linear mechanics* had considerable influence in awakening an interest in non-linear mathematics in the West.

In biology Hodgkin and Huxley interpreted the propagation of signals along a nerve in terms of a pulse of electric charge which moved along the nerve without dispersing. They invoked non-linear effects to counter the ubiquitous dispersion associated with linear waves.

In present-day jargon this is considered an example of soliton propagation. In 1957 Turing introduced a model for morphogenesis (a theory of the origin of order and life itself) based on a set of coupled non-linear chemical rate equations, and showed that the non-linearity, coupled with diffusive effects, could lead to spatially localized chemical concentrations.

Even so, until this decade, the vast majority of physicists worked and thought in terms of linear physics. A particularly compelling reason for this attitude is that quantum mechanics is very much a linear theory. The other great theory of this century, namely General Relativity, is inherently non-linear. Linearize it and you get Newtonian theory, and the geometrication of physics is reduced to a simple rigid Euclidean framework. However, this was more a matter of aesthetics, as the impact of General Relativity on particle physics (through a study of cosmological models such as the Big Bang) had not reached significant proportions. It is also fair to say that mathematical methods developed to treat the non-linear equations of this theory had not filtered down into other branches of physics.

A further branch of physics with strong non-linearity is turbulence. Long ago it was realized that turbulence was a non-linear phenomenon, but little progress towards an acceptable theory was made. Mathematical difficulties were the main reason but, perhaps, with hindsight, the lack of controlled experimental studies was also a handicap. Be that as it may, turbulence was not a popular subject. Thus it is ironic that some of the major advances made in the last few years owe much to the work carried out in this field.

Nevertheless, the past successes in physics made scientists very reluctant to turn away from their linear models. Recently a number of things have happened to change this and non-linear physics is no longer the poor relation of mainstream physics.

On the mathematical side, methods have been developed which enable one to solve exactly a class of *relevant* non-linear partial differential equations. These solutions are in the form of a set of solitons, that is, pulse-like states which propagate without change of structure, do not disperse, and can, in some cases, pass through one another without significant change: they behave more like particles than waves, although the equations themselves describe situations more usually analysed (at least in the linear sense) in terms of waves. Solitons thus become important when discussing coherent phenomena such as nerve impulse propagation and even intense heat pulses propagating in a solid.

Turbulent phenomena are usually far from being coherent and, in many cases, appear to have a purely random behaviour, such as water flowing over a weir. This behaviour can only be modelled in a linear theory by introducing a large number of individual components, just as one needs a large number of terms in a Fourier series to represent a complicated function of time. The breakthrough came with the realization that a small number of coupled *non-linear* equations can mirror many of the aspects of turbulence. The classical example is the Lorenz set of equations

$$\frac{dx}{dt} = \sigma\,(y - x)$$

$$\frac{dy}{dt} = -xz + rx - y \tag{1.3}$$

$$\frac{dz}{dt} = xy - bz\;.$$

Depending on the values of the parameters σ, r and b these equations can have time-periodic solutions but also, more importantly, solutions which are pseudo-random functions of t, that is, solutions which, although arising from completely deterministic equations, satisfy the criteria usually applied to define random numbers. A typical pseudo-random solution of the Lorenz equations is shown in Fig. 1.1, where the

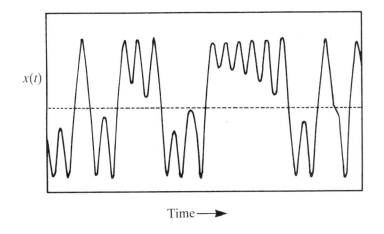

Fig. 1.1 — Variation of $x(t)$ with t for the Lorenz equation (1.3) showing typical pseudo-random behaviour.

variable x is sketched as a function of t. This behaviour is associated with a solution in the form of a *strange attractor*: a large class of different initial conditions evolve in time 'attracted' towards a unique state. This final state has limited variation in the x, y and z variables, but nevertheless undergoes time changes which never repeat themselves. This is a strange attractor, a concept introduced for the first time into mathematics by Ruelle and Takens in 1971.

The form of the strange attractor appropriate to the Lorenz equation is illustrated in Fig. 1.2, where the projection of the solution onto the y,z plane is shown. It will be noted that the trajectory spends time moving around two distinct centres. This is also apparent from Fig. 1.1, where the solution above the dotted line corresponds to motion about one centre and that below to motion about the other centre. The strangeness is associated with the seemingly random switching from one centre to the other.

A simple physical example of a strange attractor is furnished by the orbit of a satellite moving in the gravitational field of two fixed 'suns' (the restricted three body problem). Such an orbit is illustrated in Fig. 1.3, where the projection of the orbit on to a spatial plane is shown. The orbit will continue indefinitely but never return to its original position and velocity, thus making the 'ball of wool' increasingly more complicated and strange with time.

The even simpler looking difference equation, the so-called logistic equation,

$$x_{n+1} = \lambda\, x_n(1 - x_n) \ , \tag{1.4}$$

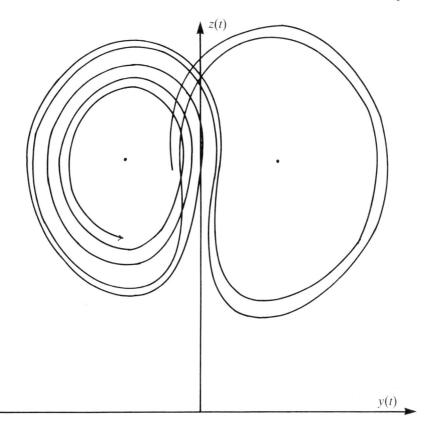

Fig. 1.2 — A plot of $z(t)$ against $y(t)$ where z and y are solutions of the Lorenz equation (1.3).
This figure illustrates the complexity of a strange attractor.

can also have periodic and non-periodic solutions as a function of n. This latter type
of solution is called *chaotic*.

The complexity of the different solutions of equation (1.4) is illustrated in Fig.
1.4, where values of x_n are plotted as functions of n for three different values of λ. For
$\lambda = 2.5$ the value of x_n approaches a unique asymptotic state whilst for $\lambda = 3.2$ the
asymptotic state is of the form of a regular alternation between two distinct values.
For $\lambda = 3.6$ no order is apparent and the behaviour is chaotic.

The knowledge that relatively simple non-linear equations can have very complicated solutions, together with the huge increase in computing power now available,
has led to enormous strides being made in our understanding of turbulent-like
behaviour.

On the experimental side, solitons have been observed in a diversity of situations
such as (a) a heat pulse propagating through a solid, (b) bunches of charge density

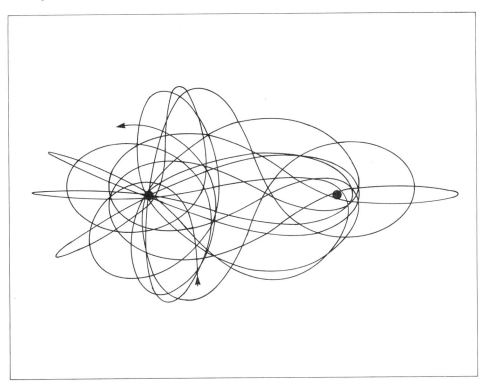

Fig. 1.3 — The orbit of a simple point mass in the gravitational field of two fixed massive bodies.

moving in a semiconductor (Gunn domains) and, of course, (c) pulses on the surface of the water. The last was the first context in which they were seen, studied and understood by Scott Russell over a hundred years ago and then totally forgotten.

Recent experiments have been carried out to study the fine detail of the onset of turbulence, and the results have been interpreted in terms of strange attractors and chaos.

Chaotic behaviour has been identified in a whole range of distinct phenomena ranging from simple (but non-linear) electrical circuits to ecological studies of population changes brought about by competition for limited resources.

Importantly, it is now realized that non-linear structures such as solitons, strange attractors and chaos are universal. For example, the Lorenz equations (1.3) model fluid turbulence where x, y and z are the amplitudes of convective structures and r the Rayleigh number. They also model the single-mode laser, with x, y and z now referring to the electric field intensity, polarization and population inversion, and r an effective pumping parameter.

Moreover, just as many different kinds of linear wave motion are describable in terms of the wave equation

$$\frac{1}{c^2} \frac{\partial^2 \phi}{\partial t^2} = \frac{\partial^2 \phi}{\partial x^2} \, ,$$

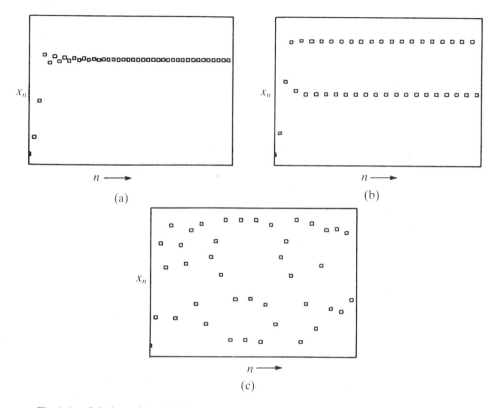

Fig. 1.4 — Solutions of the logistic equation (1.4) for different values of the parameter λ (a) $\lambda = 2.5$ (b) $\lambda = 3.2$ and (c) $\lambda = 3.6$.

many distinct types of long-wavelength disturbances satisfy the equation

$$\frac{\partial \phi}{\partial t} + \phi \frac{\partial \phi}{\partial x} + a\frac{\partial^3 \phi}{\partial x^3} = 0 \ . \tag{1.5}$$

The constants c and a contain the physics, that is, distinguish one particular phenomenon from another.

Equation (1.5) is the Korteweg–de Vries (KdV) equation. First derived for water waves by Korteweg and de Vries over a hundred years ago, it was then ignored until 1965 when Zabusky and Kruskal showed it had soliton solutions with remarkable stability properties.

These properties are illustrated in Fig. 1.5 where numerically obtained solutions of equation (1.5) are shown as a function of space for five different times. Notice how the solitons re-emerge unscathed after a collision.

In 1967 Gardner, Greene, Kruskal and Miura showed how to solve this equation

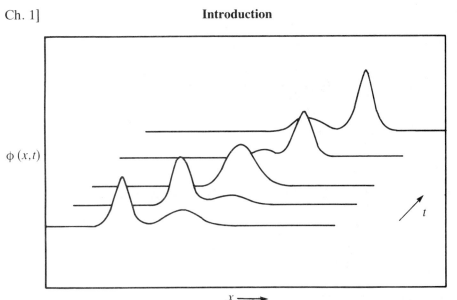

$\phi(x,t)$

Fig. 1.5 — The solutions of the KdV equation as a function of space at five distinct times. The larger soliton overtakes the slower and smaller one, and merges into a featureless solution, but then re-emerges unscathed at a later time.

exactly. A tour de force, considering that the KdV equation is a non-linear partial differential equation, but considered then by many to be a one-off method.

Any small-amplitude wave in a one-dimensional system can be expressed as a product of an amplitude and a factor $\exp(-iwt + ikx)$. Weak non-linearities are expected not only to produce a harmonic structure, but also to modulate the amplitude. For many distinct physical systems, this modulation may be described by considering the amplitude Ψ to satisfy the so-called non-linear Schrödinger equation (NLS)

$$-i\frac{\partial\Psi}{\partial t} = \frac{\partial^2\Psi}{\partial x^2} + \beta\,|\Psi|^2\,\Psi\ . \tag{1.6}$$

The particular application determines the value of β.

An important effect which has been found in a number of diverse physical situations ranging from plasma physics to fluid mechanics is where a simple modulated pulse breaks into two separate ones. Such behaviour can be studied using the NLS and is associated with the instability of a single pulse-like solution of this equation. The evolution of such an instability is illustrated in Fig. 1.6. Ocean waves satisfy this equation and it is speculated that an explanation of the famous seventh wave could come from a study of (1.6).

When in 1972 Zakharov and Shabat showed that (1.6) could also be solved exactly, the flood gates were opened. Not only were there non-linear equations with a universal range of application, but they could be solved analytically. Furthermore,

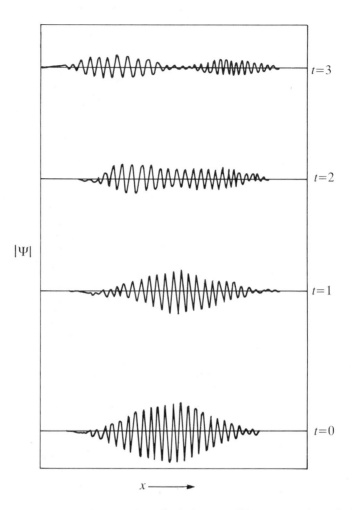

Fig. 1.6 — The break-up of a simple modulated pulse into two. This process can be studied using
the non-linear Schrödinger equation.

the solutions were in the form of solitons. This new branch of mathematical physics
had to be taken seriously.

A difficulty in studying non-linear aspects of physics is that the model equations,
such as (1.5) and (1.6), have to be derived from a more basic set of equations and the
derivation is non-trivial. In other fields of application the relationship between
model and reality is much easier to appreciate. Take, for example, from the field of
ecology, the predator–prey problem. Let $F(t)$ be the number of foxes and $R(t)$ the
number of rabbits at time t. (Since we are going to consider a differential model it is
more correct to consider R and F as the equivalent biomasses.) A reasonable
assumption, since foxes eat rabbits, is that the growth rate of rabbits decreases with
the number of foxes whilst the death rate of foxes, who are dependent entirely on the

presence of rabbits, decreases with the number of rabbits. These considerations lead to the Volterra–Lotka equations

$$\frac{dR}{dt} = (\lambda - \alpha F)R \ ,$$

and

$$\frac{dF}{dt} = -(\eta - \beta R)F \ , \tag{1.7}$$

where α, β, λ and η are all positive quantities. These simple equations capture the essence of what is found in nature, namely a periodic time variation of the populations. This periodic behaviour is illustrated in Fig. 1.7 where $R(t)$ and $F(t)$ are

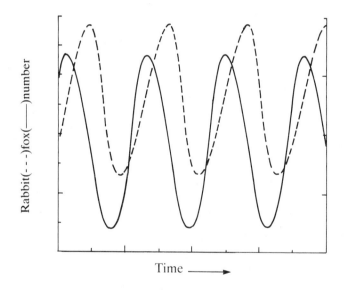

Fig. 1.7 — The time evolution of rabbit and fox populations obtained by solving the Volterra–Lotka equations (1.7). Note how the two populations, though periodic with the same period, are out of phase.

plotted as functions of time. Note that the two populations are out of phase. This reflects the biological fact built into the model that the growth rate of foxes is highest when the number of rabbits is largest.

The logistic equation (1.4) has an immediate interpretation in terms of the evolution of biological species with non-overlapping gestation periods (hence the difference rather than the differential form) subject to a self-limited growth: x_n is the number after n generations.

An experiment which is relatively easy to carry out but which may not make one popular with one's neighbours is to place a few flies or blue-bottles into a jar with a reasonable supply of meat and count their number every day. Typical results are shown in Fig. 1.8. The two sets of results correspond to two different sizes of jar, the

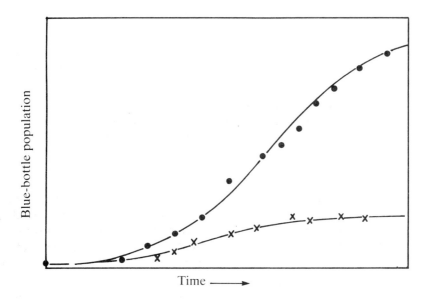

Fig. 1.8 — Population growth of flies placed in a jar as a function of time. The two separate sets of points correspond to two different sizes of jars. The upper set corresponds to the larger jar.

smaller jar giving rise, not unexpectedly, to the smaller final population. Both curves can be fitted to a solution of the logistic equation (1.4) but with different values of λ, both, however, less than three.

It could be argued that such a model, where the right-hand side of the equation is just a parabolic function of x_n, is an over-simplification. However, the exciting result of recent work is that there is a universality in the properties of such difference equations even if the right-hand side is replaced by a functional form which only qualitatively resembles a parabola. Fortunately for the progress of science the complexity of solution of such simple equations as (1.4) has been compensated for by a universality in their solutions.

A test tube of certain chemicals, free from any external stimuli, will change colour from red to blue periodically with time and continue to do so for many periods. This is an example of a Belousov–Zhabotinsky reaction, named after its two discoverers. It is the result of chemical reactions between chemicals which are blue in colour and others which are red. Consider two chemicals of concentrations C_1 and C_2

immersed in a background of other chemicals, which, for simplicity, are assumed to have constant concentrations. Chemical 1 is assumed to be produced at a constant rate α and to interact with the background chemicals to produce chemical 2 at a rate βC_1 and two atoms of chemical 1 to interact with one of chemical 2 to produce chemical 1 at a rate $\gamma C_1^2 C_2$.

Both chemicals are also assumed to diffuse in space with diffusion coefficients D_1 and D_2 respectively. In one spatial dimension this whole process is described in terms of the equations

$$\frac{\partial C_1}{\partial t} = \alpha - \beta C_1 + \gamma C_1^2 C_2 + D_1 \frac{\partial^2 C_1}{\partial x^2} \ , \tag{1.8}$$

and

$$\frac{\partial C_2}{\partial t} = \beta C_1 - \gamma C_1^2 C_2 + D_2 \frac{\partial^2 C_2}{\partial x^2} \ .$$

This is an example of a reaction–diffusion set of equations, and this particular set was first studied by a school of research centred in Brussels; hence the model is called the Brusselator. For a range of values of the parameters these equations have solutions which evolve with time from a state with no spatial structure to one which is spatially periodic. An example of such a structure is shown in Fig. 1.9. It is really surprising that such simple-looking model equations as (1.8) can have such beautiful but complicated solutions which mirror what happens in the real world. It was on the existence of such solutions that Turing in 1952 based his theory of morphogenesis, chemical separation being a precondition for cell differentiation.

Much of the understanding of the properties of non-linear equations has come from the study of dynamical systems, that is, systems evolving under the laws of Newton. At first sight this subject seems to have nothing to do with the solution of equations such as the KdV equation. However, suppose we try to find a solution to this equation, which is periodic in space and hence expandable as a Fourier series. Then writing

$$\phi(x,t) = \sum_n \phi_n(t) e^{in2\pi x/L} \ ,$$

such that $\phi(x,t) = \phi(x + L,t)$, substituting into (1.5), multiplying by $\exp(im2\pi x/L)$ and integrating over x from zero to L gives

$$\frac{d\phi_m}{dt} = i\, a \left(\frac{2\pi m}{L}\right)^2 \phi_m + i \sum_n (m + n)\, \phi_n\, \phi_{-n-m} \ ,$$

Fig. 1.9 — A typical structure that evolves from a spatially uniform state when chemicals undergo a Belousov–Zhabotinsky reaction. The shading is used to show regions where one of the chemicals has an enhanced concentration.

or more compactly

$$\frac{d\phi_m}{dt} = F_m(\phi_n) \ .$$

(1.9)

These equations may be interpreted as the equations of motion of a particle in an infinite-dimensional space with ϕ_n representing the distance from the origin in the nth direction. The time and spatial variation of $\phi(x,t)$ is thus related to a particle orbit in this space, a so-called phase space. Hence the link with dynamical systems. This link is particularly useful in obtaining an intuitive feel for the solution of non-linear partial differential equations. It may also be noted that the Lorenz equations (1.3) are of this general form. Furthermore, by writing $\theta = \phi_1$ and $d\theta/dt = \phi_2$, the simple pendulum (1.1) can be written as

$$\frac{d\phi_1}{dt} = \phi_2$$

$$\frac{d\phi_2}{dt} = -\omega^2 \sin\phi_1 \ ,$$

again of the form of (1.9).

It may also have been noted that the various mathematical models introduced above are all expressed in a small number of dependent variables; in the KdV it is just one whilst in the Lorenz system just three. This is surprising. They describe systems which are composed of many subsystems (these subsystems can be atoms) and yet they are adequately described by a few quantities. Taking over a concept from the theory of phase transformations, these quantities are called order parameters. In the theory of superconductivity the order parameter is the number of Cooper pairs whilst for a single-mode laser described by the Lorenz equations (1.3) there are three, the electric field strength and polarization and the population inversion density.

The concept of an order parameter is discussed in the next chapter with its relevance to phase transformations and Thom's catastrophe theory. If one accepts this concept, the problem of solving non-linear equations arises. In Chapters 3 and 4 methods are discussed for solving ordinary, but non-linear, differential equations of first and second order. It is true to say that solving non-linear equations is usually much harder than solving linear ones. However, a major point emerges in that the character of the solution of these equations can be obtained without actually solving them explicitly. This attitude to analysis was first appreciated by Poincaré and led him to introduce a new branch of mathematics, namely topology.

There still remains the problem of why the order parameter is a good concept. The short answer is that all the subsystems which make up the complete system are enslaved to the order parameter — just as the children of Hamlyn were enslaved to the piper. If one wanted to know the whereabouts of the children, but not a particular child, it was good enough to know the position of the piper. The mathematical justification of the enslaving principle is discussed in Chapter 5.

Putting together the concepts of an order parameter and enslaving, one can begin to understand how a system composed of many subsystems can organize itself and behave in a manner totally different from that expected of the subsystems. Importantly this organization comes about because of the interaction between the subsytems and is not due to external forces. The system self-organizes. The non-linear interactions are crucial since without them one would not get enslaving. If the children of Hamlyn had had their ears plugged it would have been a totally different story. This phenomenon of self-organization led Haken to coin the word *synergetics* for the study of such systems.

The reasons why equations such as the KdV, (1.5), and the NLS, (1.6), have such a wide range of applicability is demonstrated in Chapter 6. Here a method for their solution, the inverse scattering method, is discussed and the solution so obtained used to illustrate the unique stability of solitons.

It came as a surprise to most people that an equation as simple as the logistic equation (1.4) could have anything but simple solutions. This is far from the real situation, and once May, in 1976, had drawn attention to this, a veritable cottage industry of solving difference equations arose. One reason for this is that they are admirably suitable for numerical solution by computers.

One-dimensional difference equations such as (1.4) are discussed in Chapter 7. The form of their solutions, and most importantly their universal feature, are considered.

Just as equation (1.1) serves as a continuum model for many distinct periodic systems, the discrete set of equations

$$p_{n+1} = p_n + K\sin\theta_n \ ,$$
$$\theta_{n+1} = \theta_n + p_{n+1} \ , \hspace{4cm} (1.9)$$

serve as the universal model for systems where the dependent variables are only known at discrete intervals. For example, the above models the motion of a ball bouncing on a floor whose height changes sinusoidally with time. Then p_n is the maximum height reached by the ball and θ_n the time to reach this height, measured after n periods of oscillation of the floor. This set of equations is called the standard map or sometimes the Chirikov map, after Chirikov who first appreciated their universality. In the continuum limit where p_{n+1} may be replaced by $p_n + dp/dn$, these equations reduce to (1.1). However, unlike (1.1) they possess infinitely more varied types of solutions. In particular as the value of the parameter K is increased, the solutions change from those resembling the simple pendulum to ones which are totally chaotic and are best represented in some statistical manner. Such equations are the subject of Chapter 8.

2

Phase transformations and catastrophe theory

As was mentioned in Chapter 1 and as will emerge from the discussion of the enslaving principle in Chapter 5, the essential behaviour of complicated many-component systems can be encapsulated in the properties of an order parameter (or a small number of them). One of the simplest examples is that of a magnetic medium such as iron. The system can be described in terms of its net (average) magnetization, M, say. What can be said about such quantities from a macroscopic point of view?

The Landau theory of phase transitions rests on the assumption that a system in thermodynamic equilibrium is describable in terms of a free energy F which is a simple function of an order parameter such as M. This free energy depends also on other thermodynamic quantities, such as temperature T or pressure P. Furthermore the state of the system that exists in practice is the one where F is a minimum with respect to changes in M. That is

$$\frac{\partial F}{\partial M} = 0 \ , \qquad \frac{\partial^2 F}{\partial M^2} > 0 \ . \tag{2.1}$$

Systems in thermodynamic equilibrium are necessarily closed, that is, isolated from the rest of the world. In particular there is no input or output of energy to the system. Although many important situations are of a thermodynamic nature many more are not. The most striking ones, perhaps, are biological ones, where disordered initial systems evolve into ordered ones such as cell production or, in fact, life itself. Such phenomena are the antitheses of thermodynamics where the drive is to reduce order, that is, increase entropy. Such systems, whose behaviour depends intimately on the injection of components (chemical compounds) or energy from outside, are called open systems.

A simple but beautiful example is the Belousov–Zhabotinsky reaction where a homogeneous mixture of certain chemicals changes colour in a perfectly periodic manner, at one time red, at another blue. (The change in colour is due to the periodic change in the density of chemicals which are red and blue and whose density changes are anti-phase.) This is all done at the expense of one of the constituent chemicals, whose density slowly decreases, leading eventually to the end of the process. Order, in the form of periodic behaviour in time, is obtained at the expense of an inflow of chemicals.

The Thom theory of elementary catastrophes is based on very similar ideas to the

Landau theory but is applicable to a much wider range of systems than just thermodynamic. It was shown in Chapter 1 how a non-linear partial differential equation could be expressed in terms of a dynamical system described by a set of equations of the form

$$\frac{dA_n}{dt} = F_n(A_m) \ .$$

An important sub-class of such systems is where the 'force' F can be expressed as the gradient of a 'potential' $V(A_m)$ such that $F_n(A_m) = -\partial V/\partial A_n$. Then the equilibrium state where $dA_n/dt \equiv 0$ is such that the potential V is a minimum.

In catastrophe theory it is assumed that a potential function V exists, which is a function of state variables q_i (or order parameters) and control parameters α_i (equivalent, for example, to temperature in thermodynamics). The system that occurs in nature is a minimum of V with respect to the qs. That is, if there is just one such quantity q, then

$$\frac{\partial V}{\partial q} = 0 \ , \qquad \frac{\partial^2 V}{\partial q^2} > 0 \ . \tag{2.2}$$

Thus catastrophe theory encompasses Landau theory. However, Landau theory is based on thermodynamics and has this theory to justify the existence of the free energy F. This is not the case for other applications of catastrophe theory. The very existence of a potential function such as V is a major assumption when the theory is applied to any particular situation.

Thom showed that there are only seven basic forms for V if the number of state variables (qs) is at most two and the number of control parameters (αs) is at most three. This result is subject to a few general requirements. In particular a slight change in the form of V (other than changes in the parameters specifying V) only produces slight changes in the properties of the system (structural stability). One of the simplest of these forms, the so-called cusp catastrophe, is in fact identical to the Landau theory. Although identical, the interpretations of these theories by the physicists (Landau theory) and the mathematicians (catastrophe theory) are somewhat different. By considering both much is to be gained.

We start with the Landau theory, where a major assumption is that the free energy F can be expanded as a Taylor series in the order parameter M, so that

$$F = F_0 + F_1 M + F_2 M^2 + F_3 M^3 + F_4 M^4 \ . \tag{2.3}$$

Higher-order terms in M are neglected. The allowed values of M are those that satisfy (2.1). We insist that $M = 0$ is an extremum of F although not necessarily corresponding to a minimum. Thus $F_1 = 0$. (This can always be brought about by a linear transformation of M.)

In many systems general symmetry requirements impose the condition that $\pm M$ are equally valid solutions. In the case of a magnetic material this simply corresponds to the fact that a bar magnet is equally likely to become magnetized in either direction along its axis in the absence of an applied magnetic field. This restriction means we must take $F_3 \equiv 0$. Then the possible values of M corresponding to extrema of F are $M = 0$ and $M^2 = -F_2/2F_4$.

Now the quantities F_n depend on the temperature T. For temperatures near a critical temperature T_c, (later to be identified with the temperature where a phase transformation takes place), we expand F_2 and F_4 in powers of $T - T_c$. To lowest significant order we take

$$F_2 = \alpha(T - T_c) \quad \text{and} \quad F_4 = \beta \ ,$$

where α and β are positive quantities. This is not a restrictive as it appears at first sight. Different signs for α and/or β can be accommodated by a simple change of scale. What is important is that F_2 can undergo a change of sign as T changes from a value above T_c to one below, but F_4 cannot. Then the allowed values of M are simply

$$M = 0 \ , \quad \text{for} \quad T > T_c$$

and $M = \pm \sqrt{\alpha(T_c - T)/2\beta}$, for $T < T_c$ since M is real.

It is readily checked that the solution $M = 0$ for $T < T_c$ corresponds to a maximum not a minimum. This whole behaviour is illustrated in Fig. 2.1 where the possible

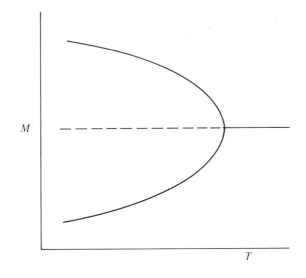

Fig. 2.1 — The variation of the order parameter M with the control parameter T. States corresponding to a minimum of the free energy are shown as continuous lines whilst the dashed line corresponds to a maximum.

values of M are shown as functions of T. The dashed line corresponds to the solution $M = 0$ for $T < T_c$, which is thermodynamically unstable ($\partial^2 F/\partial M^2 < 0$). The other values, shown as full lines, correspond to stable configurations.

Such a diagram is called a bifurcation diagram for the obvious reason that the stable state, $M = 0$, for $T > T_c$, breaks into two as T passes through the critical value T_c. This is an example of a pitchfork bifurcation because of the presence of the unstable state $M = 0$ for $T < T_c$.

In the physics literature the change as T passes through T_c is called a phase change. The phases correspond to the zero value of M ($T > T_c$) and the non-zero value for $T < T_c$. For a magnetic material the phase corresponding to $M = 0$ is called the paramagnetic phase, whilst for $M = \pm \sqrt{\alpha(T_c - T)/2\beta}$ it is called the ferromagnetic phase. The $\pm$ phases correspond to the fact that in the absence of an applied external magnetic field there is a degeneracy in direction. The critical temperature T_c in this case is called the Curie temperature.

The bifurcation of solutions is a much more general concept than that which occurs in systems in thermodynamic equilibrium. It is a concept which we shall discuss in a number of different guises later in this chapter and later in the book.

Further, it is an example of a symmetry-breaking transition, since for $T > T_c$ there is no preferred direction whilst for $T < T_c$ there is: either parallel to the direction of magnetization or antiparallel. Such transitions are much in vogue in elementary particle physics and cosmology where the role of the temperature, T, is replaced by the strength of the particle interaction. In the early universe the particle interaction was strong and neutrons and protons could be treated as if identical (symmetrical in neutron–proton interchange just as the paramagnetic is symmetrical in space). Later a phase change took place and the neutron and proton could no longer be treated the same (symmetry broken).

Returning now to the Landau theory, it is instructive to plot the free energy F as a function of M. This is done in Fig. 2.2.

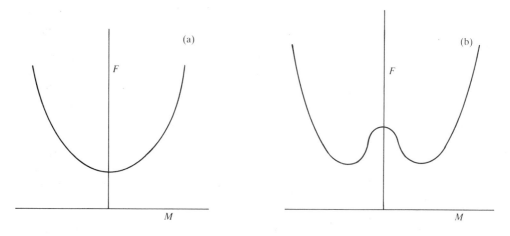

Fig. 2.2 — The variation of the free energy F with the order parameter M appropriate to a second-order phase transition (a) $T > T_c$ and (b) $T < T_c$.

For $T > T_c$, F has a single minimum at $M = 0$, whilst for $T < T_c$ there are two equivalent minima at $M = \pm M_0 (M_0 = \sqrt{\alpha(T_c - T)/2\beta})$ and now $M = 0$ is a maximum. If one considers the temperature T being gradually reduced from a value above T_c then F changes from the form shown in Fig. 2.2(a) ($T > T_c$) to that shown in Fig. 2.2(b) ($T < T_c$). Now imagine a ball in the single minimum that exists for $T > T_c$, representing the state of the system. It will remain in the minimum at $M = 0$ until T

passes through T_c, in which case it will fall into *one* of the minima at $\pm M_0$ and remain in that minimum for any further reduction in T. The change in M_0 with T is continuous throughout this process and, for historical reasons, such a phase change is called second-order.

Importantly, the thermodynamic theory cannot predict which one of the minima is selected in the transition. Fluctuations play the dominant role in such transitions and these will be discussed in the next chapter.

If the symmetry condition imposed on the above analysis is removed (if in the magnetic case an external magnetic field is applied) then F_3 is no longer zero. The free energy now takes the form

$$F = F_0 + \alpha(T - T_c)M^2 + \gamma M^3 + \beta M^4 \ , \tag{2.4}$$

where, without loss of generality, we may take α, γ and β to be positive.

For $T > T_c$, $M = 0$ is still a minimum but if $9\gamma^2 > 32\alpha\beta(T - T_c)$, a further minimum (and a maximum) now exists. Thus even for $T > T_c$ there now may be two distinct minima. This is illustrated in Fig. 2.3 where F is sketched as a function of M for a number of different values of T.

If one now imagines the ball being placed in the minimum at $M = 0$, for $T > T_c$, then it will remain in this minimum until $T = T_c$ even though another minimum exists and for a range of values of T, this other minimum has the lower value of the free energy. At $T = T_c$ the transition to this other minimum occurs and the ball remains in this minimum as T is further decreased. Now imagine the reverse process where the temperature T is increased. The transfer from one minimum to the other does not occur at $T = T_c$ but a value of T such that $9\gamma^2 = 32\alpha\beta(T - T_c)$. This is the temperature at which the one minimum ceases to exist.

It will be noted that now the state of the system, (the value of M) depends on its history, in particular whether the state is reached by increasing the temperature or by decreasing it. This phenomenon is called **hysteresis**, a name which originates from the theory of magnetism but has a much wider field of application. A hysteresis curve, corresponding to the changes taking place and illustrated in Fig. 2.3, is shown in Fig. 2.4.

A delightful illustration of hysteresis, not from physics but nevertheless physical, is given in Fig. 6.14 of Haken (1978).

Unlike the second-order phase transition discussed above ($F_3 \equiv 0$) it is now seen that M changes discontinuously as T passes through T_c. Such a transition is called first-order.

In the above we have followed a physics-based discussion of phase transitions in terms of the free energy F, the order parameter M and the temperature T. It is instructive to discuss the same type of phenomena in terms of catastrophe theory. Thus we introduce a potential V which is a function of a state variable q and two control parameters u and v. Then catastrophe theory states that the most general form for V is

$$V = \frac{q^4}{4} + \frac{uq^2}{2} + vq \ , \tag{2.5}$$

and that the system is found in the state corresponding to a minimum of V. The

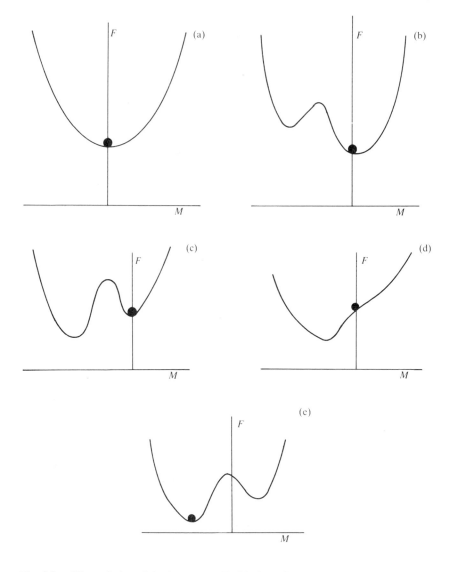

Fig. 2.3 — The variation of the free energy F with the order parameter M appropriate to a first-order phase transition. (a) $T > T_c$, $\gamma_c > \gamma$, (b) $T > T_c$, $\gamma > \sqrt{9/8}\,\gamma_c$, (c) $T > T_c$ $\sqrt{9/8}\,\gamma_c > \gamma > \gamma_c$, (d) $T = T_c$, $\gamma_c = 0$ and (e) $T < T_c$. Here $\gamma_c^2 = 32\alpha\beta(T - T_c)/9$.

variation of V with q is qualitatively the same as that of F with M, as for a first-order phase transition, and as shown in Fig. 2.3. ($M = q - 4F_3/F_4$).

From the point of view of catastrophe theory a second-order transition ($v \equiv 0$) is excluded because it is not generic. That is, such transitions are expected to be unusual since the introduction of a small non-zero value for v changes the whole phenomenon: $V(q)$ with $v \equiv 0$ is considered to be structurally unstable. It is a basic

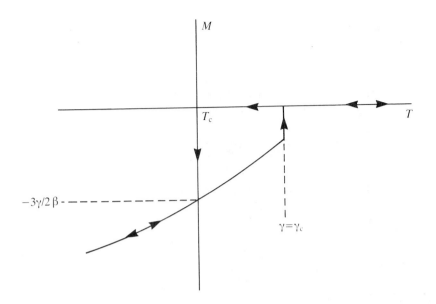

Fig. 2.4 — An illustration of the hysteresis associated with a first-order phase transition. Possible physical changes must follow the arrows. The critical value $M_c = -3\gamma/2\beta$ corresponds to the value of M at the minimum of the free energy for $T = T_c$.

requirement of the form of V is that it is structurally stable and then any information obtained from its use should be generic or universal. However, from a physics point of view this requirement is subservient to any symmetry condition and hence catastrophe theory can sometimes be over-restrictive. In the present case it excludes second-order phase transitions.

Another difference between catastrophe theory (as applied by mathematicians) and Landau type theories, is that the former pays more attention to the values of the control parameters u, v rather than the order parameter or state variable q. Thus one considers the critical lines in v, u space which separate regions where V has one minimum, from those where two exist. The condition $\partial V/\partial q = 0$ gives $q^3 + uq + v = 0$. (It is now seen why the seemingly ad hoc factors $\frac{1}{4}$ and $\frac{1}{2}$ were introduced in the definition of V.) This cubic equation for q is monotonic unless $u < 0$. For $u < 0$ the minimum of $\partial V/dq$ is at $q = \sqrt{|u|/3}$ and this touches the axis ($\partial V/dq = 0$) if $v^2 = 4|u|^3/27$. Thus for values of v larger than this critical value only one minimum exists, so that the lines $v = \pm 2|u|^{3/2}/3\sqrt{3}$ separate the v, u plane into regions of one and two minima. (These lines correspond to the condition $9\gamma^2 = 32\alpha\beta(T - T_c)$ in the Landau theory.)

It was mentioned above that where two minima exist simultaneously the history of the system dictates which minimum the system actually is in. This is the phenomenon of hysteresis. In the discussion above it was assumed that the state (or imaginary ball) would stay in the minimum in which it found itself until that minimum ceased to exist. Only then would it change. However, in practice it is found, unless one is extremely careful in an experiment, that the change between minima occurs before the actual disappearance of the second minimum. The cause of all this is that fluctuations of some form or other are ever-present. (The careful experiments alluded to above correspond to the case where these fluctuations are reduced to a minimum.) Fluctuations can push a system from one minimum to another over the intervening maximum. If the second minimum has a lower value of V, then the process is essentially one-way since the effective barrier for the reverse process is higher.

The usual way of introducing such effects, both in the Landau theory and in catastrophe theory, is to introduce the Maxwell convention or construction. This was originally introduced by Maxwell over a hundred years ago and applied to the liquid–vapour transition that occurs, for example, in water. Van der Vaal had already given an explicit form for the free energy which, near the critical temperature, reduces to the form given by (2.4).

The Maxwell construction is simple. It states that external influences or fluctuations such as thermal noise (although it is not necessary to state explicitly what they are), will drive the system so that it will always be in the state of the absolute minimum. Thus, in the u,v parameter space, one can add a line along which the two minima, when they exist, have the same value of V. In the particular case of V being given by (2.4), this line is simply $v = 0$ for $u < 0$. Thus the behaviour of the system is described in parameter space by the lines $v = \pm 2|u|^{3/2}/3\sqrt{3}$ and $v = 0 (u < 0)$. The first set separates the space into regions in which either one or two minima exist whilst the second determines where, according to the Maxwell construction, the phase change takes place (Fig. 2.5). It is important to note that with this construction

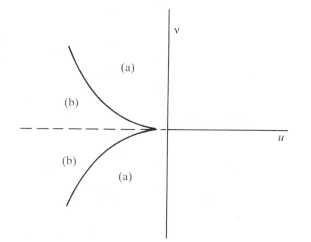

Fig. 2.5 — Parameter space plot for a cusp catastrophe. In regions labelled (a) only one minimum exists whilst in (b) two exist. The dashed line corresponds to the two minima having the same value of the free energy.

the system will no longer show hysteresis. (A second-order phase transition corresponds to the case $v \equiv 0$ and the transition taking place at $u = 0$.)

Thus for $u < 0$, an *infinitesimal* change in v will produce a finite change in the state vector q of value equal to $2\sqrt{u}$. For this reason such changes were called catastrophes by Thom. They are the equivalent in the Landau theory to a phase transition of a thermodynamic system. In particular, the above example is called a **cusp** catastrophe because of the nature of the critical lines in v, u space.

It is possible, experimentally, to reduce fluctuations to such a level as to keep the system in the unfavourable minimum and pass through the critical temperature as defined by the Maxwell construction. This is what happens for example when a liquid is supercooled below its freezing point: an increase in fluctuations at this stage leads to a rapid change of state, and this is the underlying physics governing the operation of a cloud chamber. The system is supercooled to such an extent that the passage of a single charged atomic particle is a sufficient fluctuation to initiate the rapid change of state from gas to a liquid. The track of the particle is then revealed as a track of liquid droplets.

In the case where the fluctuations are small an alternative to the Maxwell construction is needed. A common alternative is the delay convention where it is assumed that the system stays in the minimum in which it finds itself until that minimum disappears. It was this convention that was used in illustrating the phenomenon of hysteresis in a first-order phase transition and shown in Figs 2.3 and 2.4.

It is now perhaps expedient to compare the two theories. The Landau theory is based on thermodynamics and hence the existence of a free energy is assured, as well as the fact that the states of the system are associated with the minima of this quantity. The expansion of F in powers of an order parameter such as M is an assumption which in fact is known to be incorrect. However, experience has shown it to be a good approximation. Although based on thermodynamics it has a degree of universality in that it applies to a range of phenomena which includes magnetic and superconducting phase transitions.

If applied to such systems, Thom's theory gives identical results. However, the major claim is that the same type of analysis, in terms of a potential function V, which is a function of various state functions q_i and control parameters α_i, can be applied to a much wider range of phenomena. For example Zeeman has applied these ideas to a study of animal aggression, a field of study far removed from thermodynamics. In short, catastrophe theory is much more universal but the certainty of the existence of a potential function is lost. Like Landau theory, catastrophe theory also assumes that the potential can be expanded in simple powers of the qs.

For a more detailed account of catastrophe theory, including applications, see the book by Poston and Stewart (1978).

Returning now to the theory of phase transformations, it is of course important to ask how well the theory compares with the real world. The phenomenon of hysteresis, as found in the theory of first-order phase transitions, is a common feature of many diverse systems. A more quantitative comparison is obtained by considering the value of a critical exponent β defined such that for $T \rightarrow T_c$ the order parameter M is of the form $M \alpha (T_c - T)^{\beta}$. Both the Landau theory and catastrophe theory give $\beta = \frac{1}{2}$. Early experiments on many diverse physical systems suggested that this was in fact true but more exact measurements, particularly for values of T very close to T_c have shown that the value of β is of order 0.4 but still universal. This discrepancy between theory and experiment has been explained (for thermodynamic systems) by the so-called renormalization group theory (RNG), but at the expense of a considerable increase in the mathematical complexity of the theory. This theory is beyond the scope of the present book (although a simple example is discussed in Chapter 7) but one point should be made here. In RNG the free energy F is no longer taken to be an analytic function of the order parameter M, that is, F cannot be expanded as a power series in M.

The theories presented above have been uncommonly successful but it must be emphasized they still remain static theories. While they give precise information about the possible states of a system, they do not say anything about how a system may change from one state to another. Although the introduction of the Maxwell construction or the delay convention goes some way to alleviate this problem, their introduction only shows that the real problem concerns the influence of thermal fluctuations and external disturbances on the system.

These aspects are discussed in the next chapter, although it is important to emphasize that the discussion spreads from thermodynamic systems to include lasers, and biological systems where thermodynamics is not valid.

QUESTIONS

1. Obtain analytic expressions for the turning points of the free energy F as defined by (2.4). Classify them as maxima or minima.
2. Use the results obtained in question 1 to obtain the critical value of T for which the value of F is the same at the two minima.
3. Show, by using a simple linear change of variable that the general quartic potential

$$\overline{V}(p) = a_0 + a_1 p + a_2 p^2 + a_3 p^3 + a_4 p^4$$

can be expressed in the form of (2.5). Hence obtain expressions for the constants u and v in terms of the as.

4. Many years ago Van der Vaal proposed a model of a liquid–vapour phase transition which is equivalent to the following form of the free energy

$$F = (T/b)(1 - bM) \ln (1 - bM) + MT - aM^2/2$$

where T is the temperature, $1/M$ the volume and a and b constants which reflect the atomic nature of the system. Show that the above form for F reduces to the standard form of (2.4) in the limit of small M. Hence identify the critical temperature T_c in terms of a and b.

5. Use the form of F as given in question 4 to obtain an equation for the extremal values of M. Hence show that the condition for the existence of more than one extremal is $a/b > T$. Obtain an expression for T corresponding to the point of inflexion. Compare this with the value obtained by first expanding F in powers of M.

3

First-order non-linear differential equations (deterministic and stochastic)

3.1 DETERMINISTIC EQUATIONS

As was stressed in the first chapter, non-linearity holds the key to the understanding of numerous and diverse phenomena. To illustrate some of the mathematical methods that may be used to study certain non-linear features, a simple non-linear ordinary equation is discussed in some detail. Consider the equation

$$\frac{dq}{dt} = \alpha q - \beta q^3 \, , \tag{3.1}$$

where α and β are constant, and in particular β positive.

In fact such an equation describes a simple model in population dynamics: $q(t)$ is the population density at time t, α a measure of the initial growth rate, whilst the term proportional to β reflects a saturation mechanism causing a decrease in the effective growth rate with high population densities. The saturation term is not, of course, unique and many other forms, some with more biological support, have been suggested. However, here (3.1) is chosen for its simplicity.

The current, J, flowing in any medium is related to the rate of change of the electric field, E, by the simple formula

$$\frac{\partial E}{\partial t} = J \, ,$$

if all spatial variations are neglected. The current J itself depends on E and, for example, Ohm's law is nothing but the statement that $J = \sigma E$ where σ is the conductivity of the medium. For weak electric fields σ may be taken to be constant, independent of E, but this is not the case for strong fields. It is then more appropriate to write $\sigma = \sigma_0 - \sigma_1 E^2$, in which case the equation for E is of the same form as (3.1).

Equation (3.1) also describes to a good approximation the time behaviour of a single-mode laser. Here q is a measure of the intensity of the coherent radiation whilst α is a measure of the difference between the energy fed into the system in the form of incoherent radiation and the natural losses of the system (for example reflection losses at the mirrors).

Although equation (3.1) is readily solved exactly we choose not to do so, but rather use it to illustrate the essentials of linear stability analysis.

The first step in such an analysis is to obtain, if they exist, the time-independent solutions. This is very much easier than attempting to solve the equation. Such solutions satisfy (3.1) with $dq/dt \equiv 0$ and hence must satisfy

$$(\alpha - \beta \bar{q}^2)\bar{q} = 0 \ . \tag{3.2}$$

It must be stressed that these states are not necessarily thermodynamic equilibrium states. When in operation the laser, for example, is far from thermodynamic equilibrium, and biological populations can hardly be expected to be related to thermodynamic ones. However, although they are really time-independent states, it is common usage to call them equilibrium states.

One state $\bar{q} = 0$ always exists and if $\alpha > 0$, since we are taking $\beta > 0$, two further states $\bar{q} = \pm \sqrt{\alpha/\beta}$ also exist. In the biological context only one of the latter states is relevant since $\bar{q}$ must be a positive quantity. A plot of the possible values of $\bar{q}$ as a function of α is the same as the variation of M with T for a second-order phase transition as shown in Fig. 2.1; that is, we have a pitchfork bifurcation.

The simplest procedure from here is to consider the solution of (3.1) in the vicinity of the solutions $\bar{q}$. To this end we write $q(t) = \bar{q} + \delta q(t)$ and substitute into (3.1). This gives

$$\frac{d\delta q}{dt} = (\alpha\bar{q} - \beta\bar{q}^3) + (\alpha - 3\beta\bar{q}^2)\delta q + O(\delta q^2) \ . \tag{3.3}$$

All terms which involve δq^2 and higher powers are lumped together and their presence denoted by the quantity $O(\delta q^2)$. (This is a very useful shorthand notation with $O(\delta q^n)$ meaning terms of order δq^n and higher.) Also note that the first bracketed term is identically zero in view of the equilibrium condition (3.2).

The final step is to neglect all terms involving powers of δq higher than unity, that is $O(\delta q^2)$, in which case (3.3) reduces to

$$\frac{d\delta q}{dt} = \eta\delta q \ , \tag{3.4}$$

where $\eta = \alpha - 3\beta\bar{q}^2$. This is a linear equation for δq and the above type of analysis is referred to as *linearization*.

The advantage of linear equations is that, in general, they are much easier to solve than non-linear ones. In the present case (3.4) has the solution

$$\delta q(t) = \delta q_0 e^{\eta t} ,\qquad\qquad (3.5)$$

where δq_0 is an arbitrary constant of integration and, of course, corresponds to the value of $\delta q(t)$ at $t = 0$.

Now, two quite distinct types of behaviour can occur. If $\eta < 0$ then, as t increases, $\delta q(t)$ gets smaller so that $\delta q(t) \to 0$ as $t \to +\infty$. In such cases the corresponding equilibrium state is said to be **stable**. If, on the other hand, $\eta > 0$ then $\delta q(t) \to \infty$ as $t \to +\infty$ and the equilibrium is then said to be **unstable**. This means that the introduction of any small perturbation to the system ($\delta q_0 \neq 0$) will result in the solution $q(t)$ growing away from the equilibrium state.

If $\alpha < 0$ only one equilibrium state exists, $\bar{q} = 0$ and $\eta = -|\alpha|$ and the system is stable. If $\alpha > 0$ three equilibrium states exist. The state $\bar{q} = 0$ has $\eta = \alpha$ and hence is now unstable, whilst the two states $\bar{q} = \pm\sqrt{\alpha/\beta}$ have $\eta = -2\alpha$ and are hence stable. Thus the bifurcation diagram is exactly of the form shown in Fig. 2.1 with the continuous lines now corresponding to stable states and the dashed one to an unstable state.

It is important to realize the limitations of a linear analysis: if the system is stable, $\delta q(t) \leq \delta q_0$ for all t and hence, if the initial perturbation δq_0 is small, the linearization is justified for all t. This is not true for an unstable system since no matter how small δq_0, a time will be reached when $\delta q(t)$ is not small and the neglect of terms of order $\delta q^2(t)$ is no longer justified. However, it is a useful and valid procedure to *identify* unstable states. On the other hand it can give no information about the subsequent fate of the system.

It is of interest to compare the stability analysis with the Landau theory, or catastrophe theory, as discussed in the last chapter. In these latter theories it was assumed that a state could exist if $\partial V/\partial q = 0$ whilst those that actually existed satisfied the extra requirement that $\partial^2 V/\partial q^2 > 0$ (a minimum). The first of these conditions is equivalent to the condition that a time-independent state exists ($dq/dt = 0$) whilst the second is replaced by the condition of linearized stability ($\eta < 0$). This already suggests a simple way in which one can introduce dynamics into the theory presented in Chapter 2. Thus if one writes

$$\frac{dq}{dt} = -\frac{\partial V}{\partial q} ,\qquad\qquad (3.6)$$

the condition for a static solution is $\partial V/\partial\bar{q} = 0$, whilst if a linearization treatment is carried out, one gets

$$\frac{d\delta q}{dt} = \eta\,\delta q ,$$

where $\eta = -\partial^2 V/\partial \bar{q}^2$. The stability condition $\eta < 0$ is then seen to be equivalent to $\partial^2 V/\partial \bar{q}^2 > 0$. Thus, within the limits of linear theory, the dynamical theory as presented by (3.6) is equivalent to the static theory developed in the last chapter. It is for this reason that the same bifurcation diagram was obtained for both theories.

As stressed above, the linearization procedure has its limitations, particularly for unstable systems, and hence other approaches are necessary. One such approach, originally introduced by Poincaré to study non-linear second-order equations, can be used very effectively in the present case to obtain *global* information about the solution, in particular, what happens in an unstable state. To illustrate the general idea it is convenient to rewrite (3.1) in the general form

$$\frac{dq}{dt} = H(q) \equiv \alpha q - \beta q^3 . \tag{3.7}$$

The first stage is to *sketch* $H(q)$ as a function of q. This is done in Fig. 3.1(a) for $\alpha < 0$

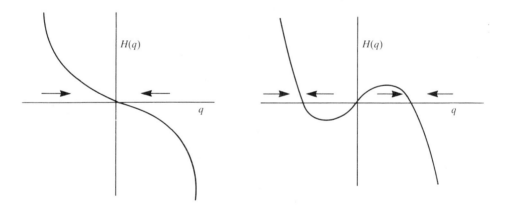

Fig. 3.1 — A phase plane analysis of a one-dimensional non-linear equation, namely equation (3.7), showing how global behaviour can be obtained without solving the equations.

and Fig. 3.1(b) for $\alpha > 0$. Now for any particular value of q one can read off the value of $dq/dt(\equiv H(q))$, and then indicate by an arrow whether q decreases or increases, that is, $H(q) < 0$ or $H(q) > 0$. For $\alpha < 0$ it is quickly seen that the arrows are always in the direction of decreasing the magnitude of q and hence the longtime solution can only approach $q = 0$ irrespective of the initial value of q. This is a simple example of an attractor. The point $q = 0$ is an attractor. For $\alpha > 0$ the arrow configuration is a little more complicated but very simple to derive. If q is initially positive, the state evolves monatomically towards $\bar{q}$ where $\bar{q}$ is the positive solution of $H(\bar{q}) = 0$, that is $\bar{q} = \sqrt{\alpha/\beta}$. Similarly, if initially q is negative, the system will evolve to the static solution $-\bar{q}$.

It is important to note that the phase plane analysis has given results with regards to any initial condition and is not limited to a linearization procedure: one can obtain global information about the solution. In the vicinity of the static equilibria $H(\bar{q}) = 0$, this information is of course consistent with that obtained from linear theory.

The static solution $\bar{q} = 0$ is an example of a repellor. In linear theory any initial condition in the vicinity of this point grows away from it exponentially with time. However, whereas the linear theory was unable to say anything about its subsequent fate, the phase plane anlysis shows that the solution can only move towards the attractors at $\bar{q} = \pm \sqrt{\alpha/\beta}$.

It will be readily appreciated that the above form of global analysis is easily extended so as to include arbitrary forms for the function $H(q)$ in (3.7). Also this global information is obtained, at most, by solving algebraic equations and not by solving differential ones.

In the next chapter, phase plane analysis will be extended to include second-order non-linear differential equations. There it will be found necessary to combine it with linear stability analysis to obtain the most useful information. However, before carrying out this extension, the effect of adding extraneous noise to equations of the form (3.1) will be discussed.

3.2 STOCHASTIC EQUATIONS

An equation such as (3.1) is said to be deterministic, which means, given the value of q at some time t, $t = 0$ say, the equation uniquely determines the value of $q(t)$ for all subsequent times. This is most easily appreciated by noting that the solution to (3.1) is of the form

$$q^2(t) = q_0^2 \, e^{2\alpha t}/\{1 - (\beta q_0^2/\alpha)(1 - e^{2\alpha t})\}$$

where $q_0 \equiv q(t = 0)$. With q_0^2 specified, $q^2(t)$ is uniquely determined for all t. It is instructive to use this solution to verify the results obtained earlier in this chapter.

When an equation of the form (3.1) is used to model a real physical situation, such as the electric field in a piece of material, a number of approximations have been made. Some of these approximations could be removed, for example, by using an equation of the form (3.7) with H a more complicated function of q. Importantly the equation (or model) would still be deterministic. There are, however, effects which do not yield to this process. For example, the conductivity of any piece of material fluctuates owing to thermal fluctuations. Although the aim of a good experiment is always to try to reduce such effects until they are negligible (in which case an equation of the form (3.7) is appropriate) this is not always possible, or even desirable, if it means carrying out the experiment in the vicinity of the absolute zero of temperature. In experiments on the controlled use of thermonuclear power the plasma is contained in a state far from thermodynamic equilibrium, and macroscopic quantities, such as the plasma density, have a significant component which is a random function of space and time. If (3.1) is used to describe the time evolution of some animal population density, then it must be expected that the parameters α and

β, which depend partly on the environment, should contain a seemingly random component due to fluctuations in, say, the weather. In some instances the effect of the outside world is to introduce fluctuations directly into the time derivative of q.

With these ideas in mind it is natural to extend an equation such as (3.1) to include fluctuations by writing

$$\frac{dq}{dt} = \alpha(t)q - \beta(t)q^3 + F(t) .$$

(3.8)

The quantities α, β and F contain contributions which are rapidly varying functions of t. This variation is rapid compared to the general time development of q.

However, in practice little is actually known about the detailed time variation of the parameters. Their time dependence is in some sense random, and hence only information in some average sense is known, that is, we specify the time variation in some statistical sense only. This being the case, it is no longer possible to obtain a complete solution of equations such as (3.8), and even if it were possible, it would not be desirable. If the coefficients in (3.8) are known only in a statistical sense, then the solution $q(t)$ is at best only going to be available in the same statistical sense. We may well have to make do with a suitably averaged solution, not a completely deterministic one.

Equations of the form (3.8) in which the coefficients and/or the driving term F have a random element with regards to their time dependence are known as **stochastic** differential equations. Subtle difficulties can arise in the treatment of such equations in that the coefficients α, β and F are not in general continuous functions of time and hence the concept of a differential loses its meaning. However, these difficulties are swept aside at this stage and the traditional methods of the calculus used. This procedure will be briefly commented on later in the chapter.

To illustrate the main features of the approach to solving stochastic equations, the linear version of (3.8) is considered, and further, the quantity α is treated as a constant. The equation is then readily solved to give

$$q(t) = q_0 \, e^{\alpha t} + \int_0^t e^{\alpha(t-t')}F(t')dt' ,$$

where q_0 is the value of $q(t)$ at $t = 0$.

Now imagine, not just one system, but many, each with the same value of α, each subject to the same initial condition, but each with a different stochastic term which is denoted by F_s with s labelling the particular system. This will give rise to different solutions denoted by $q_s(t)$ so that

$$q_s(t) = q_0 e^{\alpha t} + \int_0^t e^{\alpha(t-t')}F_s(t')dt' .$$

(3.9)

It is now natural to consider these systems to be represented by a suitable average and to this end an ensemble average is introduced, denoted by $\langle q(t) \rangle$ and defined by

$$\langle q(t) \rangle = \lim_{N \to \infty} \frac{1}{N} \sum_{s=1}^{N} q_s(t) \ . \tag{3.10}$$

Taking the ensemble average of (3.9) gives

$$\langle q(t) \rangle = q_0 e^{\alpha t} + \int_0^t e^{\alpha(t-t')} \langle F(t') \rangle dt' \ , \tag{3.11}$$

which shows that one no longer needs the $F_s(t)$s but merely their ensemble average to obtain $\langle q(t) \rangle$. Without loss of generality one can take $\langle F \rangle \equiv 0$ since a non-zero value could always be incorporated in the non-stochastic part of the equation. Thus

$$\langle q(t) \rangle = q_0 \, e^{\alpha t} \ ,$$

that is, the ensemble averaged value of q satisfies the deterministic equation.

Now, in analogy with ordinary statistical methods, one considers a measure of the deviation from the average and defines a correlation function

$$C(t,t') = \langle (q(t) - \langle q(t) \rangle)(q(t') - \langle q(t') \rangle) \rangle \ .$$

Using (3.9) one finds

$$C(t,t') = e^{\alpha(t+t')} \int_0^t dt'' \int_0^{t'} dt''' \ e^{-\alpha(t''+t''')} \langle F(t'')F(t''') \rangle \ . \tag{3.12}$$

The functions $F_s(t)$ are randomly varying functions of time, with zero mean, so that $F_s(t'')F_s(t''')$ is equally likely to be positive or negative so that the ensemble average could be zero, the exception being the case where $t'' = t'''$. Thus it is not unreasonable to consider

$$\langle F(t'')F(t''') \rangle = \sigma^2 \, \delta(t'' - t''') \ , \tag{3.13}$$

where $\delta(t)$ is the Dirac delta function and σ a measure of the variation of F_s about the mean. The above can only be taken as a plausibility argument for the statistical nature of F. Mathematically it is a very convenient form, and moreover, it has stood the test of time in that it gives results in agreement with a huge host of distinct experiments. For these reasons it is the only statistical model used in this book. Here the fluctuations (usually called noise) are, in this form, called white noise.

It is now straightforward to evaluate the integrals in (3.12) for $t \geqslant t'$

$$C(t,t') = \frac{\sigma^2}{2\alpha}\{e^{\alpha(t+t')} - e^{\alpha(t-t')}\} \ .$$

If the analysis is restricted to stable media ($\alpha < 0$) and t is taken sufficiently large, so that the effect of initial conditions are averaged away ($t, t' \to +\infty$, $t - t'$ finite) then

$$C(t,t') = (\sigma^2/2|\alpha|)e^{-|\alpha|(t-t')} \ .$$

$C(t,t')$ is a measure of the fluctuations in q_s and it is seen that these fluctuations are amplified by the factor $1/2\alpha$ above the fluctuations of F_s as measured by σ^2. In particular, this amplification is most pronounced near the critical point where α changes sign, that is, near the condition for a phase transition.

This phenomenon is actually observed, for example, in water near its freezing point. The fluctuations in density, modelled here by $\langle q^2 \rangle$, are sufficiently large to change the light-scattering properties of the water so that it appears cloudy. This is known as critical opalescence.

The above procedure could be continued to give higher moments or averages of q_s, but it is much more worthwhile to take a more direct statistical approach. Thus instead of trying to obtain the value of $q(t)$ as a function of time one asks about the probability of finding a particular value of q at a time t. We define $P(q,t)dq$ as the probability of finding $q(t)$ in the interval q to $q + dq$ at time t. It is important to realize that now q and t are treated as independent variables whereas previously q was a dependent variable dependent on t. Once $P(q,t)$ is known one can define average quantities in the usual manner, and, in particular, the average value of q, which we denote by $\bar{q}(t)$, is given by

$$\bar{q}(t) = \int q \, P(q,t) dq \ . \tag{3.14}$$

Thus $\bar{q}(t)$ is in general a function of time and, as is shown below, is equivalent to $\langle q(t) \rangle$. The probability function is normalized to unity so that

$$\int P(q,t) dq = 1 \ , \tag{3.15}$$

for all t and the integration is over all q.

Consider N systems, each of which satisfies a stochastic equation

$$\frac{dq_s}{dt} = H(q_s) + F_s(t) \ , \tag{3.16}$$

where s serves to label each member of the ensemble. $H(q)$ is a given function of q and is considered not to have any random component. For simplicity all the randomness is in the functions F_s and these are assumed to be independent of q. The probability function is defined such that

$$P(q,t) = \lim_{N \to \infty} \frac{1}{N} \sum_s^N \delta(q - q_s(t)) \ . \tag{3.17}$$

The justification is that each term in the summation is the contribution from the solutions of equation (3.16), each weighted equally. This form automatically satisfies the normalization condition (3.15) and using (3.14) one readily finds that

$$\overline{q}(t) = \lim_{N \to \infty} \frac{1}{N} \sum_s^N q_s \equiv \langle q \rangle ,$$

from the definition (3.10).

Differentiating (3.17) with respect to t and making use of (3.16) allows one to write

$$\frac{\partial P}{\partial t} = S_1 + S_2 \ ,$$

where

$$S_1 = - \lim_{N \to \infty} \frac{1}{N} \sum_s^N \delta'(q - q_s)H(q_s) \ ,$$

and

$$S_2 = - \lim_{N \to \infty} \frac{1}{N} \sum_s^N \delta'(q - q_s)F_s(t) \ .$$

$\delta'(x)$ denotes the derivative with respect to the argument. By noting the identity $f(x)\delta(x-a) \equiv f(a)\,\delta(x-a)$ for any function $f(x)$, the following simplification can be made

$$S_1 = -\frac{\partial}{\partial q}\lim_{N\to\infty}\frac{1}{N}\sum_s^N \delta(q-q_s)\,H(q_s)$$

$$= -\frac{\partial}{\partial q}\lim_{N\to\infty}\frac{H(q)}{N}\sum_s^N \delta(q-q_s)$$

$$= -\frac{\partial}{\partial q}(H(q)\,P(q,t))\ .$$

The treatment of S_2 is unfortunately more difficult. The random force $F_s(t)$ is treated as small and (3.16) solved by iteration. Thus

$$q_s(t) = \overline{q}(t) + \delta q_s(t)\ ,$$

where

$$\frac{d\overline{q}}{dt} = H(\overline{q})\ ,$$

and

$$\frac{d\delta q_s}{dt} - H'(\overline{q})\delta q_s(t) = F_s(t)\ .$$

The last equation may be solved to give

$$\delta q_s(t) = e^{\beta(t)}\int_0^t F_s(t')e^{-\beta(t')}dt'\ ,$$

where

$$\beta(t) = \int_0^t \frac{dH}{d\overline{q}}\, dt \;.$$

Insertion of this approximation into the expression for S_2 gives to lowest significant order

$$S_2 = \lim_{N \to \infty} \frac{1}{N} \sum_s^N F_s(t)\, \{\delta'(q - \overline{q}(t)) - \delta''(q - \overline{q}(t))\delta q_s(t)\} \;.$$

Taking the forcing term F_s to satisfy white noise statistics, shows that the first term in the above is zero and hence

$$S_2 = \lim_{N \to \infty} \frac{1}{N} \sum_s^N \delta''(q - \overline{q}(t)) e^{\beta(t)} \int_0^t e^{-\beta(t')}\, F_s(t) F_s(t')\, dt'$$

$$= \delta''(q - \overline{q}(t))\, \sigma^2\, e^{\beta(t)} \int_0^t e^{-\beta(t')}\, \delta(t - t')\, dt'$$

$$= \delta''(q - \overline{q}(t))\, \frac{\sigma^2}{2} \;.$$

The factor $\frac{1}{2}$ arises because the integration is only to the centre of the delta function, $t' = t$ and not through it. Finally, from the definition of $P(q,t)$, namely (3.17), it is seen that, to the lowest order in the expansion in $F_s(t)$, $P(q,t) = \delta(q - \overline{q}(t))$, so that,

$$S_2 = \frac{\sigma^2}{2} \frac{\partial^2}{\partial q^2}\, P(q,t) \;.$$

Combining the above equations gives the final equation for the probability function, namely

$$\frac{\partial P}{\partial t} + \frac{\partial}{\partial q}\, (H(q)P) = D\, \frac{\partial^2 P}{\partial q^2} \;, \tag{3.18}$$

where $D = \sigma^2/2$. This is the Fokker–Planck equation.

The stochastic nature of the problem now enters through the terms on the right-hand side. If $H \equiv 0$, then it is seen that (3.18) is a simple diffusion equation and it was for this reason that the nomenclature D was introduced. Thus stochasticity introduces a diffusion in q space, which is not necessarily ordinary space. Thus when the above analysis is applied to ecological problems, where q is associated with the number density of elephants, say, then this diffusion measures a spreading in the elephant density with time. Furthermore the larger the stochasticity as measured by σ or D, the larger the diffusion.

On the other hand if $D \equiv 0$ then the system is entirely deterministic and, as is readily found by direct substitution, the equation has the solution

$$P(q,t) = \delta(q - \overline{q}(t)) \; ,$$

where

$$\frac{d\overline{q}}{dt} = H(\overline{q}) \; ,$$

which, of course, is nothing but (3.16) with $F_s \equiv 0$.

One of the assumptions made in the above derivation was that $F_s(t)$ was independent of q.

If it is still assumed that $F_s(t)$ is of the form of white noise, but with σ a function of q, then the above analysis still goes through, with minor modification, and one obtains an equation of the form

$$\frac{\partial P}{\partial t} + \frac{\partial}{\partial q}(HP) = \frac{\partial}{\partial q}\left(D(q)\frac{\partial P}{\partial q}\right) , \qquad (3.19)$$

Integration of both forms of the Fokker–Planck equation over all q, subject to the reasonable condition that P tends to zero as $q \to \pm \infty$, shows that

$$\frac{\partial}{\partial t}\int_{-\infty}^{+\infty} P(q,t)dq = 0 \; .$$

This, of course, reflects the normalization of the probability function as demanded by (3.15).

The above derivation of the Fokker–Planck equation relies heavily on the assumption that dq/dt exists even in the presence of the stochastic term. In practice this term could be extremely pathological. These difficulties can be overcome by considering the change in q over a small but finite time rather than dq/dt. Two approaches have been suggested and developed into the Îto and the Stratonovich calculus. For the case where $F_s(t)$ is independent of q they both give the same result, namely (3.18). Otherwise they differ, with the Stratonovich method giving (3.19). Experimental evidence to date, though somewhat meagre, is in favour of the Stratonovich results and hence the form of the Fokker–Planck equation given above.

The removal of the other restrictions in the above proof, such as the restriction to white noise, can be made, but only at the expense of a considerable increase in the complexity of the final equation. Since an equation of the form (3.18) cannot be solved in general, it is difficult to justify the introduction of further complication unless absolutely essential. Thus here we adopt the view that (3.18) or (3.19) are sufficiently good model equations for most applications.

Unfortunately the Fokker–Planck equation is not very amenable to analysis. However, by considering two particular solutions, some of the features of the general solution are illuminated.

The first class of solutions is those which are time-independent. Then equation (3.19) with $\partial P/\partial t \equiv 0$ is readily integrated to give

$$P(q) = P_0 \, e^{\int g (H(q')/D(q'))\mathrm{d}q'},$$

where P_0 is the normalization constant, and by insisting that $P \to 0$ as $q \to \pm \infty$ the second integration constant has been set to zero. Using the particular form for $H(q)$ implied by (3.8), and taking $F(t)$ to be independent of q, allows the above to be written as

$$P(q) = P_0 e^{-V(q)/\sigma^2}, \tag{3.20}$$

where $V(q) = -\alpha q^2 + \beta q^4/2$.

This result is very reminiscent of the Maxwell–Boltzmann distribution found in the statistical mechanical treatment of interacting atomic systems. Then V is the total energy of the system and $\sigma \equiv k_B T$, where T is the temperature and k_B Boltzmann's constant. However (3.20) is much more general. It applies to systems which are not necessarily in thermodynamic equilibrium, such as the laser, and even to elephant populations where the whole concept of thermodynamic equilibrium is meaningless. (What is the temperature associated with a group of elephants?)

It is now interesting to compare the solution (3.20) with the steady state (stable) obtained in the deterministic case ($F \equiv 0$). For $\alpha < 0$ the general form for $P(q)$ is as sketched in Fig. 3.2(a) where the width is of order σ and the height of order $1/\sigma$. In the limit $\sigma \to 0$, $P(q)$ reduces to a delta function about $q = 0$, which is of course the deterministic result, namely just a single equilibrium state at $q = 0$. For $\alpha > 0$, $P(q)$ is as sketched in Fig. 3.2(b). There are now two equivalent maxima centred about $q = \pm \sqrt{\alpha/\beta}$ and a minimum at $q \equiv 0$. In the limit $\sigma \to 0$ the distribution shrinks to two delta functions centred about $q = \pm \sqrt{\alpha/\beta}$. These states are just the stable states found in the deterministic case. The unstable state found in this case corresponds to the minimum of P at $q = 0$.

There is a vital difference between the stochastic and the deterministic cases. For the latter, if $q(t=0) > 0$ then $q(t) \to +\sqrt{\alpha/\beta}$ for large times ($q(t=0) < 0$ gives $q(t) \to -\sqrt{\alpha/\beta}$), whereas when randomness is introduced, no matter how small,

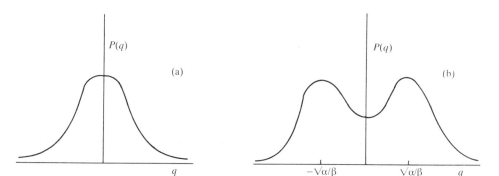

Fig. 3.2 — Sketch of the time-independent probability function $P(q)$ as a function of q for (a)
$\alpha < 0$ and (b) $\alpha > 0$.

there is a finite probability that the system will exist with either value $\pm \sqrt{a/\beta}$, no matter what the initial value of q. There is also a finite probability that the system will exist in the state $q = 0$, which is an unstable state for the purely deterministic situation for $\alpha > 0$.

The above results all make good physical sense, and though there may be doubts about the derivation of the Fokker–Planck equation and its range of validity, it seems to encapsulate much of the physics.

The full time-dependent solution of the Fokker–Planck is notoriously difficult to obtain analytically for general functions $H(q)$. However, the flavour of this general solution is captured by the special case where both H and D are constants. The equation is then of the form of heat conduction (P is equivalent to the temperature) in a medium which is being convected with a velocity equivalent to H. For the particular case where $P(q,t=0) = \delta(q - q_0)$ one finds

$$P(q,t) = (1/(\pi\eta(t)^{1/2})e^{-(q-\bar{q}(t))^2/\eta(t)} \ ,$$

where $\bar{q}(t) = q_0 e^{-Ht}$ and

$$\eta(t) = (D/2H)(1 - e^{-2Ht}) \ .$$

Putting $H \equiv 0$ gives $\eta = Dt$ and $\bar{q} = q_0$. Then the above equation for P shows the well-known relaxation of a pulse with time, familiar from the analogy with a heat pulse. For H finite, η approaches a constant for large t, so that the broadening of the pulse also approaches a limit. The centre of the pulse now moves away from q_0 towards the origin. Thus, whereas D is responsible for the diffusive behaviour, a finite value of H produces a drift in the probability density as a function of q, as well as significantly altering the magnitude of the diffusion.

A quantity of considerable physical interest, which depends on the solution of the

Fokker–Planck equations, but which itself is much easier to calculate than the full solution, is the mean first-passage time M. Consider the Fokker–Planck equation valid in a region $q_1 < q < q_2$, and subject to the boundary conditions that $P(q,t)$ is identically zero at these boundaries for all t. That is, $P(q_1,t) = P(q_2,t) = 0$. Physically this corresponds to the presence of perfectly absorbing boundaries at q_1 and q_2. For simplicity it is also assumed that at $t = t_0$, $P(q,t) = \delta(q - q_0)$.

To make this initial condition explicit it is convenient to write the solution of the equation in the form $P(q,t;q_0,t_0)$. Now the probability that q never reaches the boundaries in the time interval t_0 to t is simply

$$W(t;q_0,t_0) = \int_{q_1}^{q_2} P(q,t;q_0,t_0) \, dq \ .$$

The probability that q reaches either boundary in the time interval t to $t + dt$ is equal to $\dfrac{d}{dt}(1 - W) \, dt$, so that the average or mean first-passage time is given by

$$M(q_0) = \int_{t_0}^{\infty} t \frac{d}{dt}(1 - W) \, dt,$$

which by using the above definition of W, reduces to

$$M(q_0) = \int_{t_0}^{\infty} dt \int_{q_1}^{q_2} P(q,t;q_0,t_0) dq \ . \tag{3.21}$$

It would seem necessary to solve the equation for P before obtaining $M(q_0)$. Fortunately this is not the case, and, as shown in Appendix A.1, $M(q)$ itself satisfies the ordinary differential equation

$$D \frac{d^2 M}{dq^2} + H(q) \frac{dM}{dq} = -1 \ , \tag{3.23}$$

subject to the boundary conditions that $M(q_1) = M(q_2) = 0$. Note that this is in fact a first-order differential equation for dM/dq, so it is easily integrated once to give

$$\frac{dM}{dq} = -\frac{1}{D} e^{-V(q)} \int^{q} e^{V(q')} \, dq' \ ,$$

where

$$V(q) = +\frac{1}{D} \int^q H(q')dq' \ .$$

The equation can be integrated once more and the evaluation of $M(q)$ reduced to quadrature. In general, little progress can be made analytically except for the special case where

$$H(q) = +h_0 \ q>0$$
$$= -h_0 \ q<0 \ ,$$

so that $V(q) = h_0|q|/D$. Then for the symmetric case $q_2 = -q_1$, and for $q_0 \geqslant 0$

$$M(q_0) = \frac{D}{h_0^2}\left\{(\bar{q}_1 - \bar{q}_0) - (e^{-\bar{q}_0} - e^{-\bar{q}_1})\right\} \ ,$$

where $\bar{q} = h_0q/D$.

QUESTIONS

1 Use equation (3.1) to obtain a differential equation for $\phi = q^2$. Solve this equation for ϕ.

2. Use the stability analysis developed in section (3.1) to study the equation

$$\frac{dq}{dt} = \alpha q - \beta q^2$$

where α, β are independent of t and both positive.

3. Use the analysis developed in section (3.1) to study the equation

$$\frac{dq}{dt} = \sin q$$

4. Show that the exact solution of the equation given in question 3 is

$$q = 2 \tan^{-1} (\tan(q_0/2)e^t)$$

where q_0 is the value of q at $t = 0$. Use this result to study the global features of the differential equation.

5. Write down the Fokker–Planck equation appropriate to the deterministic equation

$$\frac{dq}{dt} = -a \tanh q$$

Obtain the full solution of the time-independent Fokker–Planck for the case where the diffusion coefficient is a constant. For what value of q is this probability function a maximum? How does this compare with the solution of the deterministic equation?

6. Use equation (3.23) to calculate the mean first-passage time M for the particular case where $D = \frac{1}{2}$ and $H(q) = \tanh q$ with $M = 0$ for $q = \pm 1$.

4

Second-order non-linear differential equations and phase plane analysis

4.1 INTRODUCTION

Probably the first non-linear differential equation an aspiring young physicist comes across is the one describing the simple pendulum. If θ is the angular displacement, l the distance between the point of suspension and the centre of gravity of the bob, and g the acceleration due to gravity, then

$$\frac{d^2\theta}{dt^2} + (g/l)\sin\theta = 0 \ . \tag{4.1}$$

This equation is difficult to solve in general, but, if attention is restricted to small displacements, $\sin\theta$ may be approximated by θ and the above equation then reduces to the well-known equation for a simple harmonic oscillator, namely

$$\frac{d^2\theta}{dt^2} + \omega^2\theta = 0 \ , \tag{4.2}$$

where $\omega^2 = g/l$. This equation, as is well-known, has solutions which are periodic, period $2\pi/\omega$. A substantial part of physics is based on this simple equation, from the phonons which propagate in solids, to the waves on the surface of the oceans. The linearized version of (4.1), namely (4.2), encapsulates so much of the behaviour of (4.1), that there has been a reluctance to consider the full non-linear phenomenon. Further, experiments are, if possible, designed to ensure that the linear regime is applicable. This is not always possible and one of the aims of this book is to show that such a procedure is not always necessary. Moreover, many fascinating phenomena can *only* be understood by taking into account the non-linearity.

The simple pendulum is a good example. Equation (4.1) may be integrated once to give

$$\frac{1}{2}\left(\frac{d\theta}{dt}\right)^2 = C + \omega^2\cos\theta \ , \tag{4.3}$$

where C is a constant of integration which is essentially the energy of the system. Now this equation may be integrated once more, and θ obtained as an explicit function of t, in terms of Jacobian elliptic functions. Though these functions are in many ways similar to trigonometric functions, in that they are periodic functions, they have a secondary complexity which reduces their appeal compared to the trigonometric functions. Fortunately all the main properties of the solutions of (4.1) can be obtained by simple algebraic considerations applied to (4.3). In Fig. 4.1(a) the

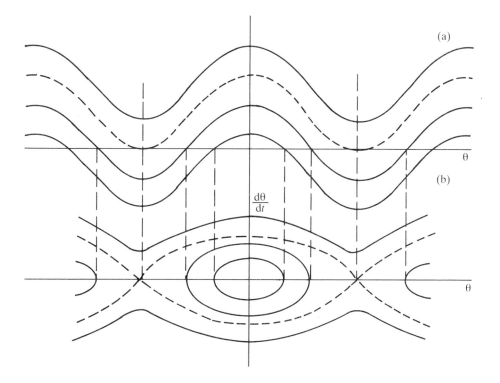

Fig. 4.1 — The phase plane portrait corresponding to equation (4.3). In (a) the right-hand side of (4.3) is shown as a function of θ for four different values of the constant C. The corresponding variation of $d\theta/dt$ with θ is shown in (b). The special case $C = \omega^2$ is shown as a dashed curve and in (b) this dashed curve is the separatrix.

right-hand side of (4.3) is sketched as a function of θ for four distinct values of C. By using these curves together with (4.3) the variation of dθ/dt with θ is readily obtained and this variation is sketched in Fig. 4.1.(b). There are three distinct types of curve:

(a) Closed curves which enclose the origin (θ = dθ/dt = 0); these correspond to solutions of (4.1) where the pendulum oscillates periodically in time but always such that $|\theta| < \pi$. The two curves correspond to different values of C, that is, to different initial conditions for the solutions of (4.1). They also have different periods.

(b) Curves which correspond to the pendulum making complete circles of revolution; here dθ/dt is always of one sign. One curve is for clockwise rotation, the other for an anti-clockwise motion. We are, of course considering the pendulum to be made with a rigid arm.

(c) A critical curve which corresponds to the pendulum being released from a vertical position(θ = − π) passing through zero and slowly reaching the vertical position (θ = + π) again. In practice, this particular solution is never achieved, since, as will be shown later, it takes an infinite time to go from θ = − π to θ = + π. However, this curve is extremely important as it separates solutions for which θ remains bounded (− π < θ < π) from those where θ can take all values. Such a curve is called a separatrix.

In summary, all the main features of pendulum motion are represented by the curves in Fig. 4.1(b). This information was obtained without knowing the explicit solutions to (4.1). Curves were sketched rather than differential equations solved.

Fig. 4.1(b) represents an example of a phase plane. A natural question to ask is whether these ideas can be generalized to other non-linear differential equations. This is the question that phase plane analysis (or the qualitative theory of differential equations) attempts to answer.

4.2   PHASE PLANE ANALYSIS

It is convenient to express the basic equations in the form of two coupled first-order equations

$$\frac{dx}{dt} = H(x,y) \ , \tag{4.4}$$

and

$$\frac{dy}{dt} = R(x,y) \ . \tag{4.5}$$

H and R are arbitrary functions of x and y. If H and R do not depend explicitly on t then the equations are called autonomous. Otherwise they are said to be nonautonomous. Here we consider the autonomous case only. Equation (4.1) is readily expressed in the above form with $\theta \equiv x$, $d\theta/dt \equiv y$, $H \equiv y$ and $R \equiv -\omega^2\sin x$.

Equations (4.4) and (4.5) are similar to those describing a single particle subject to some external force (x equivalent to position and y to momentum), that is, they are similar to equations describing a dynamical system, namely Newton's equations. Though not strictly of Newtonian form the solution is often described in the language of particle motion. Thus the solution of (4.4) and (4.5), as a function of t, is described as the trajectory or orbit (of a particle) moving through x,y phase space.

From (4.4) and (4.5) we have

$$\frac{dy}{dx} = R(x,y)/H(x,y) \tag{4.6}$$

and hence the tangent (dy/dx) to the trajectory in the x,y plane is uniquely defined, except at critical points $\bar{x},\bar{y}$, where $H(\bar{x},\bar{y}) = R(\bar{x},\bar{y}) = 0$. Thus, except possibly at these critical points, trajectories in the x,y plane cannot cross. It is extremely important to remember this seemingly trivial result when one attempts to sketch the global behaviour of trajectories in the phase plane.

From the physical point of view the critical points $\bar{x},\bar{y}$ are nothing but the equilibrium or time-dependent solutions of (4.4) and (4.5). They are sometimes called fixed points.

Conceding that it is usually too difficult to solve (4.4) and (4.5), it is reasonable, as a first step at least, to consider the solution in the neighbourhood of the fixed points. Thus we write $x(t) = \bar{x} + \delta x(t)$, $y(t) = \bar{y} + \delta y(t)$, substitute into (4.4) and (4.5), express the right-hand sides of these equations as Taylor series in terms of δx and δy, and neglect products of these small quantities. That is, we linearize the equations. This gives

$$\frac{d\delta x}{dt} = \alpha\, \delta x + \beta\delta y \ ,$$

$$\frac{d\delta y}{dt} = \gamma\delta x + \eta\delta y \ ,$$

where the quantities α, β, γ, η are independent of t, but depend on the form of H and R near $x = \bar{x}$ and $y = \bar{y}$. For example $\alpha = \partial H/\partial x$ evaluated at the critical point. Such equations are readily solved in terms of exponentials and their solution is

$$\delta x(t) = Ae^{\lambda_+ t} + Be^{\lambda_- t} \ , \tag{4.7}$$

$$\delta y(t) = (\gamma A/(\lambda_+ - \eta))e^{\lambda_+ t} + (\gamma B/(\lambda_- - \eta))e^{\delta_- t} \ , \tag{4.8}$$

where $\lambda_{\pm}$ satisfies the so-called indicial equation

$$\lambda^2 - (\alpha + \eta)\lambda + \alpha\eta - \beta\gamma = 0 \ , \tag{4.9}$$

so that

$$2\lambda_{\pm} = \alpha + \eta \pm \sqrt{(\alpha - \eta)^2 + 4\beta\gamma} \ . \tag{4.10}$$

The above analysis is identical to a linear stability analysis of the equilibrium specified by $x = \bar{x}$, $y = \bar{y}$.

The constants A and B serve to specify the initial values of δx and δy at $t = 0$. The behaviour of $x(t)$ and $y(t)$ in the vicinity of a fixed point is now specified uniquely in terms of the above expressions for δx and δy, and we use this information to classify the different types of possible behaviour.

(a) *Stable node (simple point attractor)*

If λ_+ and λ_- are both real and negative then as $t \to \infty$, $\delta x \to 0$ and $\delta y \to 0$ for all values of A and B. Thus all trajectories approach, and finally end up at the fixed point. This behaviour is illustrated in Fig. 4.2(a) with the arrows denoting the

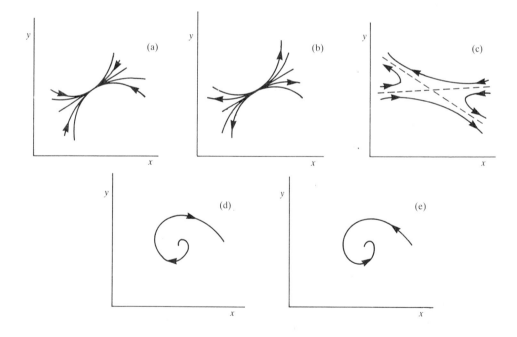

Fig. 4.2 — Phase trajectories near (a) stable node, (b) unstable node, (c) a saddle point, (d) unstable focus, and (e) a stable focus. In (c) the dashed lines correspond to the separatrix. The arrows indicate the change in solution with a positive change in time.

direction of time. The different trajectories correspond to different initial conditions, that is, different values of A and B.

(b) *Unstable node (simple point repellor)*

Here λ_+ and λ_- are still real but now both positive. All trajectories move away from the fixed point. This is illustrated in Fig. 4.2(b).

(c) *Saddle point*

Here λ_+ and λ_- are real but of opposite sign. We take $\lambda_+ > 0$.

Now for $t \to +\infty$, the term proportional to B in δx and δy, may be neglected, and it is then seen that all the trajectories asymptotically approach the straight line given by $\delta y = (\gamma/(\lambda_+ - \eta))\delta x$. Similarly for $t \to -\infty$ all trajectories approach the line $\delta y = (\gamma/(\lambda_- - \eta))\delta x$. Note that the slopes of these lines depend on the properties of H and R but *not* on the initial values of δx and δy. These two lines form the **separatrix**. Since trajectories cannot cross, except possibly at fixed points, the separatrix divides the phase plane into four distinct regions. The form of possible trajectories can now be easily sketched and is shown in Fig. 4.2(c) with the dashed line representing the separatrix.

Examples of saddle points are given in Fig. 4.1(b). They are the points $(\pm\pi,0)$. The tangents to the dashed curve, at these points, form the separatrix.

The trajectory that lies on the separatrix and moves towards the fixed point as t increases corresponds to the special case $A \equiv 0$. Then $\delta x = B\exp(-|\lambda_-|t)$ so that the trajectory takes an infinite time to reach the fixed point ($\delta x = 0$). Thus the dashed curve in Fig. 4.1(b) corresponds to periodic motion with an infinite period.

(d) *Unstable focus*

When the roots of the indicial equation are complex, (4.7) and (4.8) may be rewritten as

$$\delta x = A e^{st} \cos(\omega t + \phi)$$

and

$$\delta y = \frac{A e^{st}}{\beta} \{(s - \alpha)\cos(\omega t + \phi) - \omega\sin(\omega t + \phi)\}$$

where $\lambda_\pm = s \pm i\omega$. The trajectories now have the form of spirals and for $s > 0$ they spiral out, away from the fixed point, as illustrated in Fig. 4.2(d).

(e) *Stable focus*

Here $s < 0$ and the trajectories spiral into the fixed point as shown in Fig. 4.2(e).

(f) *Centre*

This is the special case of $s \equiv 0$, in which case the trajectories are in the form of closed curves. They correspond to periodic solutions, period $2\pi/\omega$. A good example of this type of behaviour is the simple pendulum, as illustrated in Fig. 4.1, with the point $(0,0)$ the centre (although only contours near this centre are describable by the linear theory).

A corollary to this result is that if a closed trajectory exists in the phase plane, then it must enclose a fixed point.

A centre is a special type of solution ($s \equiv 0$) and one can imagine that any small change to the forms of H or R will lead to a finite value for s. The centre will then change to a stable or unstable focus. A well-known example of this behaviour is the effect of adding a little damping to the simple pendulum equation (4.1). The periodic solutions (centres) now decay in amplitude until eventually the pendulum comes to rest at $\theta = 0$ (stable focus). Importantly, a further change in the amount of damping does not change the qualitative features of the solution. This is an example of structural instability: a small change in the form of the equations (here in H and/or R) leads to a *qualitative* change in the solution. Such considerations have led to the requirement that sets of equations, used to model the real world, must be structurally stable so that small changes in the equations only lead to small changes in the solutions. For example, in Thom's catastrophe theory this condition is imposed on the potential.

4.3   LIMIT CYCLES (ATTRACTORS)

Because of the linearization, the above analysis and classification are only applicable in the vicinity of the fixed points.

To obtain a full understanding of the possible types of solutions of equations (4.4) and (4.5), one needs to know the global features of the phase plane. Sometimes these may be guessed from the form of the solution near the fixed points and the requirement that trajectories do not cross. Fig. 4.1(b) is a good example where the guessing is easy. However, what is possibly the most important type of solution cannot be guessed unless one has the intuition of Poincaré. This solution is the *limit cycle*. The limit cycle is a closed curve in the phase plane and hence corresponds to periodic behaviour. However, it is *not* a solution of the basic equations. Actual solutions do, however, approach the limit cycle asymptotically with time. A limit cycle is stable if neighbouring trajectories approach it as $t \rightarrow +\infty$ (an attractor), and unstable if trajectories diverge from it as $t \rightarrow +\infty$ (repellor). An illustration of a stable limit cycle is shown in Fig. 4.3).

From the point of view of modelling the real world, the differences between the trajectories associated with a centre and a limit cycle are worth emphasizing. Though both correspond to periodic solutions, the centre is structurally unstable, the limit cycle is not. Furthermore, the solution in the vicinity of a centre is uniquely determined by the initial conditions, whereas a limit cycle attracts all trajectories in its vicinity, and each one of them corresponds to different initial conditions. The limit cycle reflects the nature of the equations and not the initial conditions.

In summary, we define a stable limit cycle as a closed curve in the phase plane which all trajectories in its vicinity approach asymptotically with time.

A simple example is furnished by (4.4) and (4.5) with

$$H = x + y - x(x^2 + y^2) \ ,$$

and

$$R = -(x-y) - y(x^2 + y^2) \ .$$

The only critical point is (0.0) and a linear analysis readily shows that this point corresponds to an unstable focus ($s = +1$, $\omega = 1$). Further progress in x, y geometry is difficult to make, but, if the equations are expressed in terms of cylindrical coordinates ($x = r\cos\theta$, $y = r\sin\theta$), then (4.4) and (4.5) reduce to

$$\frac{\mathrm{d}r}{\mathrm{d}t} = r(1 - r^2) \ ,$$

and

$$\frac{\mathrm{d}\theta}{\mathrm{d}t} = -1 \ .$$

These equations are readily integrated to give $\theta = \theta_0 - t$ and

$$r^2 = \frac{Ae^{2t}}{(1 + Ae^{2t})} \ ,$$

where A and θ_0 are determined by the initial conditions. Typical trajectories are shown in Fig. 4.3. These are of the form of spirals which for large t approach the circle

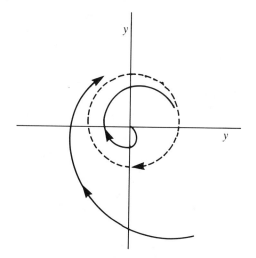

Fig. 4.3 — A simple example of a limit cycle generated by the equations $\mathrm{d}r/\mathrm{d}t = r(1 - r^2)$ and $\mathrm{d}\theta/\mathrm{d}t = -1$ with $x = r\cos\theta$ and $y = r\sin\theta$. The limit cycle is the circle $r = 1$.

$r = 1$. This is the limit cycle. The above form for r^2 shows that for $t \to +\infty$, $r^2 \to 1$, independently of the value of the initial conditions (A, θ_0). For $A > 0$ the initial point is inside the limit cycle and the trajectories spiral out to the circle $r = 1$, whilst for $A < 0$ the initial point is outside the limit cycle and trajectories spiral inwards.

As may have been guessed by now the above example is somewhat contrived. The equation was first considered in r, θ space and back-transformed to x, y space. However a non-trivial example is the Van der Pol equation

$$\frac{d^2 x}{dt^2} - a(1 - x^2) \frac{dx}{dt} + \omega_0^2 x = 0 , \tag{4.11}$$

which differs from the simple linear pendulum equation by the inclusion of a non-linear damping. The equation arises in the study of simple electric circuits which contain a non-ohmic resistance. In fact, such circuits were used by Van der Pol to give the first experimental verification of the concept of a limit cycle.

This equation can be put into standard form with $y = dx/dt$, $H = y$ and $R = -\omega_0 x + a(1 - x^2) y$. The origin $x = y = 0$ is the only fixed point, and linear analysis shows this point to be a focus with $s = a/2$ and $\omega^2 = \omega_0^2 - a^2/4$ ($a^2/4 < \omega_0^2$). Thus, for $a > 0$, the origin is an unstable focus.

The form of (4.11) suggests that the local growth rate (local value of s) is $a(1 - x^2)$ so that for $x^2 > 1$ one may expect trajectories to converge towards the origin. This suggests the possibility that a limit cycle exists. Numerical computations confirm these suspicions and a typical solution is shown in Fig. 4.4. (Physically the existence of the limit cycle means that after a short time, sufficient to allow transients (initial conditions) to die away, the system oscillates periodically with time.)

To study this problem analytically it is necessary to resort to some form of perturbation theory. For small values of a/ω_0, the time scale of the oscillatory motion $(2\pi/\omega_0)$ is small compared to the time scale associated with the expansion of the orbit $(1/a)$. The spiralling motion is tight. A powerful perturbative method, based on the existence of two distinct time scales, and called the multiple-time perturbation method, is discussed in Appendix A2. To apply this method to (4.11) we introduce the expansion parameter ε and write

$$x(t) = x_0(\tau_0, \tau_1) + \varepsilon \, x_1(\tau_0, \tau_1) + O(\varepsilon^2) ,$$

with $\tau_n = \varepsilon^n t$ and further order a to be of order ε; that is, write $a = \varepsilon a$. Then, proceeding as in Appendix A2, one obtains a solution

$$x(t) = A\cos(\omega_0 t + \phi) - (aA^3/32\omega_0)\sin(3(\omega_0 t + \phi)) + O(A^5),$$

where $\tag{4.12}$

$$\frac{dA}{dt} = \frac{aA}{2} (1 - A^2/4)$$

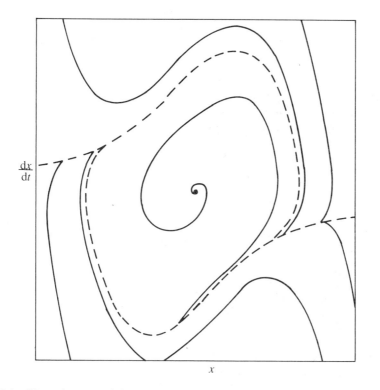

$\dfrac{\mathrm{d}x}{\mathrm{d}t}$

x

Fig. 4.4 — Phase plane portrait for the Van der Pol oscillator (equation (4.11) with $a = \omega_0 = 1$) showing the presence of a limit cycle shown as a dashed curve.

and ϕ is constant. The real expansion parameter is now seen to be aA^2/ω_0. This solution shows all the characteristics of trajectories in the vicinity of a limit cycle. The limit cycle itself is obtained from the condition that $\mathrm{d}A/\mathrm{d}t = 0$, that is, $A^2 = 4$. The solution for $x(t)$ is then periodic. For small values of a the limit cycle is near circular, but it gets increasingly distorted as the value of a is increased. Fig. 4.4 shows the limit cycle obtained numerically for $a = 1$.

4.4   THE EFFECTIVE POTENTIAL

In a significant number of applications the basic equations arise, not in the form of (4.4) and (4.5), but as a second-order equation

$$\frac{\mathrm{d}^2x}{\mathrm{d}t^2} = F(x) \; , \tag{4.13}$$

where $F(x)$ is a given non-linear function of x. Of course this form can be expressed in the form of (4.4) and (4.5) simply by writing $y = \mathrm{d}x/\mathrm{d}t$ and then $H \equiv y$ and $R \equiv F$. An example of this type of equation is equation (4.1). The above equation can be integrated once and expressed in the form

$$\frac{1}{2}\left(\frac{dx}{dt}\right)^2 = K - V(x) \; , \tag{4.14}$$

where K is a constant of integration and by analogy with Newtonian dynamics $V(x)$ is an effective potential given by $F(x) = -\dfrac{\partial V}{\partial x}$. Knowing $V(x)$ it is then very easy to sketch the *complete* phase plane portrait. This is illustrated in Fig. 4.5.

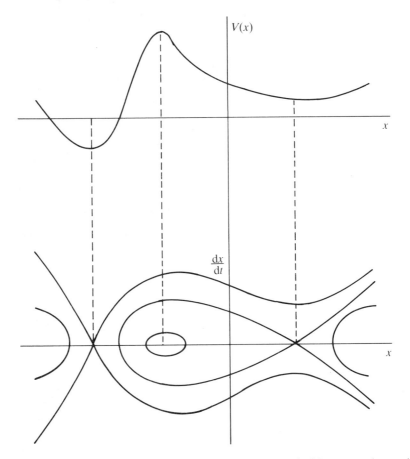

Fig. 4.5 — Illustration of how a phase plane portrait may be sketched for a general potential $V(x)$.

4.5   THE PREDATOR–PREY PROBLEM

This problem, described by equation (1.7), is an example where the whole global structure of the phase plane can be unravelled, even though it is not possible to solve the equations analytically.

The equations are of the form of (4.4) and (4.5), with the replacement $R \to x$, $F \to y$, $H(x,y) = (\lambda - \alpha y)x$ and $R(x,y) = -(\eta - \beta x)y$. There are two fixed points $(0,0)$ and $(\eta/\beta, \lambda/\alpha)$. The origin corresponds to a saddle point with the separatrices coinciding with the x and y axes. The point $(\eta/\beta, \lambda/\alpha)$ is a centre. This information, together with the requirement that the trajectories cannot cross, allows one to sketch the behaviour in the whole of the phase plane, as shown in Fig. 4.6. This leads to the

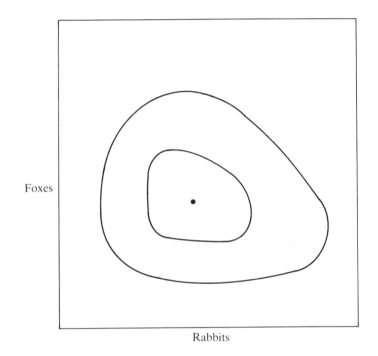

Foxes

Rabbits

Fig. 4.6 — Phase plane portrait for the predator–prey problem as described by equation (1.7) or equations (4.4) and (4.5).

important result that all solutions of the equation (1.7) in the physically acceptable part of the phase plane ($R > 0$, $F > 0$) are periodic in time. Such is the power of phase plane analysis. A limitation is that it cannot give quantitative information such as the value of the period of a solution.

The existence of a centre shows that equations (1.7) are structurally unstable, and hence of limited use as a biological model. Structurally stable equations, based on the existence of a limit cycle, have replaced (1.7) in biological modelling. One such example is where

$$H = (\lambda - \alpha y)x - kx/(x + a) \; ,$$

$$R = \eta y(1 - \beta y/x) \; ,$$

where λ, α, k, η and β are all constants.

The effect of the introduction of stochasticity into (1.7) by the addition of a random term $H_s(t)$, and a term $R_s(t)$ into (4.4) and (4.5) respectively, can be treated in a manner similar to that given in section 3.2. The appropriate Fokker–Planck equation now takes the form

$$\frac{\partial P}{\partial t}(x,y,t) + \frac{\partial}{\partial x}(H\,P) + \frac{\partial}{\partial y}(R\,P) = \frac{\partial}{\partial x}\left(D_1\frac{\partial P}{\partial x}\right) + \frac{\partial}{\partial y}\left(D_2\frac{\partial P}{\partial y}\right) ,$$

where we have assumed

$$\langle H_s(t)\,H_s(t')\rangle = \sigma_1^2\,\delta(t-t') ,$$
$$\langle R_s(t)\,R_s(t')\rangle = \sigma_2^2\,\delta(t-t') ,$$

but

$$\langle H_s(t)\,R_s(t')\rangle \equiv 0 .$$

This latter assumption removes the possibility of cross-field diffusion. Here $D_1 = \sigma_1^2/2$ and $D_2 = \sigma_2^2/2$.

It has been found, by numerical computation, that for H and R given for the predator–prey model by (1.7), the only time-independent solution of the stochastic equation is where P and F are both zero. This result confirms the structural instability of the deterministic equations (1.7). In this case the forms of the solutions of the equations are *qualitatively* different when only a small amount of stochasticity is introduced.

4.6 HIGHER-ORDER EQUATIONS

The linearization procedure introduced in this chapter to discuss the nature of the solution of second-order equations in the vicinity of fixed points can be readily extended to include higher-order equations. An indicial equation is still obtained but in general it is no longer possible to obtain the solutions in closed analytic form. However, its solutions can still only be in terms of real or complex eigenvalues and one can distinguish between attracting directions and repulsive ones. The topology or form of the solution near a fixed point is not always easy to visualize and of course one loses the two-dimensionality of the phase plane. Nevertheless, it is still the first thing to do when one is confronted with a new equation. Some of these points can be illustrated by considering the Lorenz equation (1.3).

There is a fixed point for x, y and z all zero and near this point the linearized equations take the form

$$\frac{d\delta x}{dt} = \sigma(\delta y - \delta x) \ ,$$

$$\frac{d\delta y}{dt} = r\delta x - \delta y \ ,$$

$$\frac{d\delta z}{dt} = -b\delta z \ .$$

The last equation immediately gives $\delta z = C \exp(-bt)$ where C is some constant, so that all trajectories are attracted to the origin in the z direction. The other two equations are solved by looking for solutions proportional to $\exp(\lambda t)$ in which case the condition for a non-trivial solution (the indicial equation) takes the form

$$\lambda^2 + (1 + \sigma)\lambda + \sigma(1 - r) = 0 \ .$$

For $r < 1$ both solutions are real and negative in which case the origin is a three-dimensional attractor. Physically this corresponds to the fluid having uniform density and no flow or the laser in the quiescent state (not lasing). For $r > 1$ one of the eigenvalues is real and positive in which case trajectories move away from the origin in one direction. We now have *one* possible three-dimensional extension of a saddle point. The trajectories are attracted to a fixed point along two directions and are repelled along the other direction. Physically, for $r > 1$ the fluid in the uniform state is unstable.

There are two other fixed points given by $\bar{x} = \bar{y} = \pm \sqrt{b(r-1)}$ and $\bar{z} = r - 1$ which of course only exist for $r > 1$. Linearizing the equations about these fixed points and looking for solutions proportional to $\exp(\lambda t)$ gives, as the condition for the existence of a non-trivial solution, the indicial equation

$$\lambda^3 + \lambda^2 (1 + b + \sigma) + b(r + \sigma)\lambda + 2b\sigma(r - 1) = 0 \ . \tag{4.15}$$

Since for $r > 1$ all the coefficients in the above equation are positive only one real solution exists and this is always negative. From this result follows an important consequence. All trajectories in the vicinity of these fixed points are attracted to a two-dimensional surface (manifold) and their subsequent motion is essentially two-dimensional.

To proceed further we assume the other two solutions to be purely imaginary. The substitution of $\lambda = i\omega_0$, with ω_0 real, into (4.15) followed by equating the real and imaginary parts to zero gives two equations for ω_0^2, namely

$$\omega_0^2 = b(r + \sigma) \quad \text{and} \quad \omega_0^2 = 2b\sigma(r - 1)/(1 + b + \sigma) \ .$$

For consistency these must be identical and this condition leads to a critical value of r, namely r_c, which is a function of the quantities b and σ. Simple algebra gives

$$r_c = \sigma\,(b + \sigma + 3)/(\sigma - b - 1) \ ,$$

with obvious restrictions on σ and b ($\sigma > b + 1$).

Near this critical value we write $\lambda = i\omega_0 + \eta$ and $r = r_c + \delta r$, substitute into (4.15) and retain terms of order δr only (η is complex and of order δr). This gives

$$\eta = \frac{+\,b\delta r\,(2\sigma + i\omega_0)/2\omega_0}{[\omega_0 - i(1 + b + \sigma)]}$$

which corresponds to a change in the real and imaginary part of λ. Importantly we have ($\eta = \eta_R + i\eta_I$)

$$\eta_R = b(\sigma - 1 - b)(r - r_c)/D \ ,$$

where $D = 2[\omega_0^2 + (1 + b + \sigma)^2]$.

For $r < r_c$, $\eta_R < 0$ the three-dimensional phase plane near these fixed points is of the form of a stable focus on a two-dimensional surface. This changes to an unstable focus for $r > r_c$.

Thus for $1 < r < r_c$ we conjecture that all trajectories in the three-dimensional phase plane spiral into one or other of the two stable foci at $\bar{x} = \bar{y} = \pm\sqrt{b(r - 1)}$, $\bar{z} = r - 1$.

Numerical simulations show this to be true and we have another example where the global features can be inferred from the behaviour near the fixed points.

For $r > r_c$ all the fixed points are unstable and there seems nowhere for trajectories to end. By analogy with the discussion of similar situations in the two-dimensional case one would naturally look for the three-dimensional analogue of a limit cycle. This would be in the form of a torus or sphere. The numerical simulations of Lorenz and the analysis of Ruelle and Takens showed that this was not the case. An attractor exists on which all trajectories eventually end up but this attractor is no simple geometric object such as a torus. It is called a strange attractor and its complicated form, or strangeness, may be appreciated by looking at Fig. 1.2. Here a projection of this three-dimensional object (the full solution of the Lorenz equation) onto the z, y plane is shown. The properties of strange attractors are discussed further in Chapter 8.

4.7 SUMMARY

The form of the phase plane portrait in the vicinity of a fixed point has been discussed. It is interesting to note that the mathematical analysis is identical to a linear stability analysis of a steady-state solution.

For a complete qualitative description of a non-linear system it is necessary to obtain the global features of the phase plane. Unfortunately no general method of obtaining such information exists, except detailed numerical computation. However, in some cases it is possible to infer the global behaviour from the nature of the solution in the vicinity of the fixed points. This procedure has been illustrated by considering the predator–prey problem.

A global feature of importance is a limit cycle. The nature of such a structure has been analysed in some detail by applying the multiple-time perturbation theory to the Van der Pol equation.

QUESTIONS

1. Use phase plane analysis to discuss the nature of the fixed points of the equation for the simple pendulum, equation (4.1).
2. Apply phase plane analysis to the damped simple pendulum for which a term $a \, d\theta/dt$ is added to the left-hand side of (4.1). Consider the limit $a\sqrt{l/g} \rightarrow 0$.
3. The Volterra–Lotka equations (1.7) can in fact be integrated once to give

$$\eta \, \ln R - \beta \, R - (\alpha F - \lambda \ln F) = C$$

   where C is a constant of integration. Show that this result is consistent with the equation (1.7).
4. Show that a necessary condition for the existence of a stable limit cycle for the equation

$$\frac{d^2 x}{dt^2} + f(x) \, \frac{dx}{dt} + \omega^2 \, x = 0$$

   is that $f(0) < 0$ and $4\omega^2 > f^2(0)$.
5. Use the results of section (4.4) to derive the potential corresponding to the force $F(x)$ defined such that

$$F(x) = -a \quad \text{for} \quad x > 0$$
$$= +a \quad \text{for} \quad x < 0$$

   Use equation (4.14) to obtain an explicit expression for the variation of x with t and show that the motion is periodic. Calculate the period.

5

Self-organization and enslaving

Before the Pied Piper went to Hamlyn the movement of the rats in the town was more or less random. Once he played his magic flute their movement became very ordered; they all followed the piper. The motion of the rats was enslaved to that of the piper; and their motion was essentially that of the piper. Thus one may consider the piper as an order parameter, and describe the main features of the motion of the rats in terms of this one quantity alone. This is obviously a simplification of a kind which would be most desirable in a treatment of any system composed of many subsystems. It is the aim of this chapter to illustrate how an order parameter can be identified and an equation for its time evolution obtained.

Consider the set of differential equations ($i = 1,2,3 \ldots$)

$$\frac{dq_i}{dt} = -\alpha_i q_i + A e^{-\lambda t} , \qquad (5.1)$$

where one could take the q_is to represent the rats, and the driving term, the piper. The solution for each 'rat' is

$$q_i(t) = q_{0,i} e^{-\alpha_i t} + \frac{A}{(\alpha_i - \lambda)} (e^{-\lambda t} - e^{-\alpha_i t}) ,$$

where $q_{0,i}$ is the initial value ($t = 0$) of q_i. Now suppose the characteristic times $1/\alpha_i$ are all much shorter than the time $1/\lambda$, then after sufficient time for the effects of the initial conditions to die away ($\alpha_i t \gg 1$), the above solutions reduce to

$$q_i(t) = \left(\frac{A}{\alpha_i}\right) e^{-\lambda t} .$$

That is, the time dependence of the q_is is governed by the external driving term. One can say 'the qs are enslaved to the driving term'.

Now notice that this enslaved solution may be obtained directly from (5.1) by simply putting the term $dq_i/dt = 0$. This does not mean that the q_is are independent of t but that they instantaneously relax to the equilibrium state ($dq/dt = 0$) and acquire a time dependence through that of the driving term. This is the enslaving principle. It is readily extended to the set of equations

$$\frac{dq_i}{dt} = H_i(q_j(t)) + S_i(t) \ ,$$

where the H_is are functions of the q_is and S_is are external driving terms. If all the subsystems, the q_is, interact on time scales much shorter than the characteristic time scales associated with the S_is, then the term dq_i/dt may be dropped. The q_is then come to a dynamical equilibrium described by

$$H_i(q_j(t)) = -S_i(t) \ .$$

Mathematically, the importance of this whole procedure is that the problem of solving differential equations is replaced by that of solving algebraic ones.

However, a much more interesting situation arises in systems which produce their own driving terms. These internal driving terms are then order parameters, and, hopefully, are few in number. If this is the case the subsystems have self-organized such that the properties of the system as a whole can be described in terms of a few variables: the order parameters.

To illustrate the above possibility consider the coupled set of equations

$$\frac{dq_1}{dt} = \lambda_1 q_1 - a q_1 q_2 \ , \tag{5.2}$$

$$\frac{dq_2}{dt} = -\lambda_2 q_2 + b q_1^2 \ , \tag{5.3}$$

where λ_1, λ_2, a and b are all positive constants. These equations may be analysed by the methods described in Chapter 4. Using such methods, it is found that the critical point $(0,0)$ is a saddle point with the separatrices along the coordinate axes. There are two stable foci at the points $(\pm \sqrt{\lambda_1 \lambda_2/ab}, \lambda_1/a)$ (at least for $\lambda_1 > 8\lambda_2$ which is the case of interest). The global structure of the whole phase plane can now be readily sketched. In particular, if the treatment is limited to initial values of q_1 and q_2 both positive, then all trajectories eventually end up at the focus $(+\sqrt{\lambda_1 \lambda_2/ab}, \lambda_1/a)$. In the neighbourhood of this point the trajectories are of the form

$$q_1 = \bar{q} + Ae^{-\eta_+ t} + Be^{-\eta_- t} \ ,$$

$$q_2 = \lambda_1/a - (A\eta_+/a\bar{q})e^{-\eta_+ t} - (B\eta_-/a\bar{q})e^{-\eta_- t} \ ,$$

where $\bar{q} = \sqrt{\lambda_1 \lambda_2/ab}$ and

$$2\eta_\pm = -\lambda_2 \pm \sqrt{\lambda_2^2 - 8\lambda_1 \lambda_2} \ .$$

Now if there is a disparity in time scales such that $\lambda_2 \gg \lambda_1$, then $\eta_+ \gg \eta_- \approx 2\lambda_1$ and

$$q_2 = \lambda_1/a - (\lambda_1/a\bar{q})(q_1 - \bar{q}) \ . \tag{5.4}$$

This means that in the neighbourhood of the fixed point all the trajectories are attracted to a unique straight line. On this line the actual dynamics is described by $q_1 = \bar{q} + B\exp(-2\lambda_1 t)$.

In the case where $\lambda_2 \gg \lambda_1$, the basic equations (5.2) and (5.3) may be simplified by using the enslaving principle. One envisages that q_2 relaxes quickly to a dynamic equilibrium, obtained from (5.3) by neglecting the term dq_2/dt. Thus $q_2 = bq_1^2/\lambda_2$, which, when substituted into (5.2), gives

$$\frac{dq_1}{dt} = \lambda_1 q_1 - abq_1^3/\lambda_2 \ . \tag{5.5}$$

This equation was examined in Chapter 3 where it was shown that for large times $q_1 \to \bar{q}(\pm\sqrt{\lambda_1\lambda_2/ab})$ and for values of q_1 near this value

$$q_1(t) = \bar{q} + Be^{-2\lambda_1 t} \ .$$

This result is in total agreement with the phase plane analysis, which, near the fixed point, may be considered exact. Furthermore by considering the equation for the dynamical equilibrium in the neighbourhood of the fixed points, (5.4) can be derived.

In summary, when two disparate time scales exist, $(\lambda_2 \gg \lambda_1)$, the behaviour of the system as it approaches the fixed point is fully captured by the simplified equations obtained using the enslaving principle. In this case the variable q_1 may be identified with the order parameter.

The above considerations are readily extended to the case of more than just two coupled equations in which the characteristic time of evolution of one of the components is slower than all the others.

For example the system

$$\frac{dq_1}{dt} = \lambda_1 q_1 + H_1(q_i) \ ,$$

and

$$\frac{dq_i}{dt} = -\lambda_i q_i + H_i(q_j) \ ,$$

where $\lambda_1 \ll \lambda_i$ all i. Applying the enslaving principle to all the equations except the first gives a set of algebraic equations, $q_i = +H_i(q_j)/\lambda_i$, which in principle can be solved to give $q_i = G_i(q_1)$; that is, all the q_is can be expressed in terms of the one variable q_1. This is identified as the order parameter and satisfies the equation

$$\frac{dq_1}{dt} = \lambda_1 q_1 + H_1(G_i(q_1)) \ .$$

If more than one time scale is slow compared with all the others, then the enslaving principle is used to eliminate the fast time variation and express all the fast variables in terms of the slow ones. These are now identified with order parameters.

The method of enslaving is used in many branches of physics but in different guises. Thus in solid state physics and chemistry where one examines the motion of electrons in a solid, the electrons are assumed to move in an instantaneous static lattice of ions. This is a reasonable procedure because of the disparity in mass between electrons and atoms (or ions). The atoms are sluggish, the electrons mobile and thus the sluggish atoms enslave the mobile electrons. This is known as the Born–Oppenheimer approximation. Another example is the theory of the single-mode laser. Here there are three distinct time scales: the shortest, associated with atomic transitions, an intermediate one associated with the inversion population, and the slowest, associated with the build-up of intensity in the laser. The enslaving principle is first applied to the atomic motion where the atoms are assumed to relax instantaneously to an equilibrium controlled by the time-varying inversion population and laser intensity. It is then applied to the dynamics of the inversion population such that this population relaxes to an equilibrium controlled by the laser intensity only. This intensity may then be identified with an order parameter. Detailed calculations show that the final equation for the intensity is of the form (3.1), with α a measure of the difference between the pumping rate and loss rates due to, for example, imperfect reflection at the mirrors.

Here, however, to illustrate in a non-trivial manner the application of the enslaving principle, we consider solving a reaction–diffusion type problem. As a simple example we consider a one-component system, in one spatial dimension, described by the equation

$$\frac{\partial C}{\partial t} = \alpha C - \beta C^3 + D\frac{\partial^2 C}{\partial x^2} \ . \tag{5.6}$$

The quantity C may be the chemical concentration or the density or biomass of a biological species. Its temporal behaviour is controlled by reaction processes whilst its spatial variation is controlled by diffusion. Hence the name reaction–diffusion for this type of equation. We consider the system enclosed in a box of length L with a no-flux condition applied at each end. Thus $\partial C/\partial x = 0$, for all t, at $x = 0$ and $x = L$. These boundary conditions are most readily incorporated by writing

$$C(x,t) = \sum_{-\infty}^{+\infty} A_m e^{im\pi x/L} \ ,$$

with $A_m = A_{-m}$. With the A_ms real, this condition ensures that C is real and also that the boundary conditions are automatically satisfied. Substitution of this expression into (5.6), multiplication by $\exp(-in\pi x/L)$ and subsequent integration over x from 0 to L, gives, for all n,

$$\frac{dA_n}{dt} = -\lambda_n A_n - \beta\sum_{l,m}A_l A_m A_{n-l-m} \ , \tag{5.7}$$

where $\lambda_n = D(n\pi/L)^2 - \alpha$. This is an infinite set of coupled non-linear differential equations and at this stage the introduction of the Fourier series for $C(x,t)$ seems of little advantage. Note that this set is of the form of a dynamical system discussed earlier in this book.

If $\alpha < D(\pi/L)^2$ then all the λ_ns, except $\lambda_0(= -\alpha)$ are positive and the linear part of (5.7) suggests that the A_ns will decay exponentially with time, except possibly A_0. Assuming all the A_ns to be zero $(n \geqslant 1)$, (5.7) reduces to

$$\frac{dA_0}{dt} = \alpha A_0 - \beta A_0^3 \tag{5.8}$$

This is a familiar equation with an asymptotic solution $A_0^2 = \beta/\alpha$. We can consider A_0 to be the order parameter, but this whole solution is very uninteresting since it corresponds to a spatially uniform solution.

Now suppose that $0 < \alpha < 4D(\pi/L)^2$ so that both λ_0 and λ_1 are negative but all the other λ_ns are positive. Linear theory suggests that A_0 and A_1 are more important than the other A_ns, since the latter amplitudes decay with time.

In such a case the enslaving principle can be generalized to treat these unstable modes as the order parameters, and allow all the other A_ns to be enslaved to them. Thus (5.7) is reduced to

$$\frac{dA_0}{dt} = \alpha A_0 - \beta S_0 \ ,$$

$$\frac{dA_1}{dt} = (\alpha - D(\pi/L)^2 A_1 - \beta S_1 \ ,$$

but

$$A_n = \beta S_n/\lambda_n \ ,$$

for $n \geqslant 2$. Here βS_n is the non-linear term that occurs in (5.8). Thus the number of dynamical equations has been reduced to two at the expense of having to solve an infinite set of algebraic equations. Of course it is impossible to solve these algebraic equations and so some further approximations have to be made.

The simplest is to put $A_n \equiv 0$, $n \geqslant 2$, in which case the system reduces to

$$\frac{dA_0}{dt} = \alpha A_0 - \beta A_0^3 - 6\beta A_o A_1^2 \ , \tag{5.9}$$

$$\frac{dA_1}{dt} = -\lambda_1 A_1 - 3\beta A_1(A_1^2 + A_0^2) \ . \tag{5.10}$$

Note that if we assume A_1 to be enslaved, (5.10) gives either $A_1 = 0$ or $A_1^2 = -A_0^2 - \lambda_1/3\beta$. Since the As are real, this latter solution is only possible for $\lambda_1 < 0$. Thus if $\lambda_1 > 0$ the only solution is $A_1 = 0$ and then the above reduce to (5.8). This gives some justification to the whole approach.

The above equations are of the form studied in Chapter 4 and the simple substitution $x = A_0^2$, $y = A_1^2$ reduces them to standard form (4.4), (4.5)

$$\frac{dx}{dt} = 2x(\alpha - \beta(x + 6y)) \ , \tag{5.11}$$

and

$$\frac{dy}{dt} = 2y(\lambda - 3\beta(x + y)) \ . \tag{5.12}$$

Here $\lambda = -\lambda_1$, so that $\lambda > 0$. Application of the method discussed in Chapter 4 allows the fixed points of the above equations to be classified as follows.

$(0,0)$ unstable node

$(0,\lambda/3\beta)$ stable node if $2\lambda > \alpha$
                                saddle if $2\lambda < \alpha$

$(\alpha/\beta,0)$ stable node if $3\alpha > \lambda$
                                saddle if $3\alpha < \lambda$

$$\left(\frac{2\lambda - \alpha}{5\beta},\ \frac{3\alpha - \lambda}{15\beta}\right) \text{ saddle}$$

Since $x(=A_0^2)$, $y(=A_1^2)$ are positive, we only need consider the positive quadrant of the phase plane, in which case the last saddle point only exists for $2\lambda > \alpha > \lambda/3$. Note that if initially $x = 0$, then $dx/dt = 0$, and the trajectory must be along the $x = 0$ axis. Similarly the $y = 0$ axis is a particular trajectory. Since trajectories cannot cross, the axes serve as barriers between the quadrants. The phase portraits are sketched in Fig. 5.1 for the separate cases $\alpha > 2\lambda$ and $\alpha < 2\lambda$. These sketches are readily made by

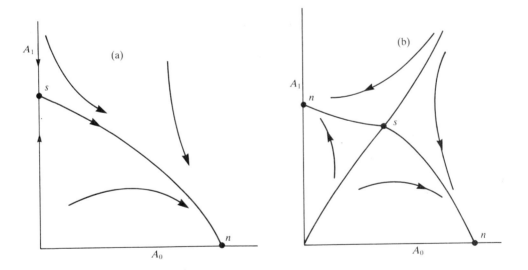

Fig. 5.1 — Phase plane portraits corresponding to the coupled equations (5.9) and (5.10) with (a) $\alpha > 2\lambda$ and (b) $6\alpha > 2\lambda > \alpha$. The letters s and n denote saddle point and stable nodes respectively.

using the known form of solution about the fixed points and remembering that trajectories cannot cross. In Fig. 5.1(a) the case $\alpha > 2\lambda$ is illustrated and it is seen that all trajectories are attached to the stable node at $(\alpha/\beta,0)$. This corresponds to

$A_0 = \sqrt{\alpha/\beta}$ and $A_1 = 0$, the uninteresting state with no spatial structure. Note that this is the eventual outcome even though $\lambda_1 < 0$ (linear instability) as long as $|\lambda_1| < \alpha/2$. For $3\alpha > \lambda > \alpha/2$ the phase portrait changes in a qualitative manner and is now as shown in Fig. 5.1(b). There are now two stable fixed points. The point to which the system will evolve depends on initial conditions, as is clear from the phase plane portrait. To generate a spatially varying structure $(A_1 \neq 0)$, the initial state must have a sufficiently large value of A_1.

Inclusion of the next term A_2, but using the enslaving principle to express it in terms of A_0 and A_1, still leads to equations of the form of (5.11) and (5.12) but with modified coefficients of β. For $\lambda_2 > 0$ the inclusion of the extra terms does not change the qualitative features of the phase portrait and hence no qualitative changes in the form of the solution are expected. This is not too unexpected, but gratifying, with regard to the consistency of the method of approximation, since $\lambda_2 > 0$ corresponds to the A_2 mode being linearly stable. For $\lambda_2 < 0$ and $|\lambda_2|$ greater than some critical value, it is expected that the introduction of a third-order parameter would be necessary. Somewhat surprisingly in the case of three order parameters, the type of solution found is chaotic, but this is the subject of later chapters and will not be pursued further here.

Although the above reaction–diffusion model, as described by (5.6), is oversimplified to have direct application to the real world, it does illustrate two important phenomena. Firstly, that a non-linear system can, without the necessity of introducing external forces, evolve with time to a non-trivial and unique final state. This is an example of self-organization. Secondly, given that the initial conditions satisfy rather mild conditions (for example y large enough), the system evolves to a state with well-defined spatial structure. This is a simple example of morphogenesis.

QUESTIONS

1. Use the exact solution of (5.5) as given in section (3.2) to show that for $t \to \infty$

   $$q_1 = \bar{q} + Be^{-2\lambda_1 t}$$

   and hence obtain values for $\bar{q}$ and B. For what initial value of q_1 can one obtain a time-independent solution to (5.5)?

2. Use the enslaving principle to obtain an approximate solution to the set of equations

   $$\frac{dq_1}{dt} = q_1 + \sqrt{(a^2 + q_2^2)}$$

   $$\frac{dq_2}{dt} = \lambda(q_2 - q_1)$$

   where $\lambda \gg 1$. Hence find the behaviour of q_1 for large values of t.

3. Show how by rescaling both the dependent and independent variables the problem described by equation (5.6) can be reduced to a similar equation involving only one constant. Hence obtain the rescaled version of equation (5.7).

4. Obtain the exact solution of the coupled set of linear equations

$$\frac{dq_1}{dt} = \lambda_1 q_1 + aq_2 \ ,$$

$$\frac{dq_2}{dt} = -\lambda_2 q_2 + bq_1 \ .$$

Consider the form of this solution in the limit $\lambda_2 \gg \lambda_1$. Compare this latter solution with one obtained directly using the enslaving principle.

6

Solitons

6.1 INTRODUCTION

Most media are dispersive: when small amplitude pulses propagate in such media, they will tend to broaden, eventually losing all spatial variation; that is, they disperse. Interestingly, in some media, if the amplitude of the pulse is increased, then the pulse will retain its shape as it propagates. In short, the dispersive nature of the medium has been balanced by non-linear effects. Such pulses are called solitary waves or solitons. Though the existence of such solutions has been known for over a century, with both theoretical and experimental justification, their importance has only been realized in the last twenty years.

A good example of a solitary pulse is the wave formed on the surface of water in a long channel (such as a canal) due to the abrupt stopping of a barge. It was in fact the study of such waves that first led Scott-Russel, as long ago as 1844, to formulate the physical basis of what are now called solitons. However, this work, and that of Korteweg and de Vries (1895), who obtained, as an approximation, the form of the equation that describes soliton properties, namely (1.5), was long forgotten. In 1965 Zabusky and Kruskal, in studying a different problem, rederived the equation of Korteweg and de Vries (now called the KdV equation), solved it numerically, and discovered that the solution, in the form of a number of solitons, had quite remarkable properties. Contrary to all previous beliefs and best opinions, these pulse shapes could pass through one another and eventually emerge unscathed. Thus, not only did they travel without dispersing, they behaved more like particles than waves when they interacted. Hence the name soliton. The next milestone in the history of solitons was passed in 1967 when Gardner, Greene, Kruskal and Miura showed how the KdV equation, a non-linear partial differential equation, could be solved *exactly*. Then in 1972 Zakharov and Shabat showed how the method could be applied to another non-linear partial differential equation (the non-linear Schröd-inger equation (1.6)). Since then a whole industry has been created for the study and application of solitons.

As was mentioned in Chapter 1, the concept of a soliton has been applied to a vast and diverse range of scientific situations, from high temperature plasma physics to muscle contraction in animals and humans.

The name soliton strictly applies to non-linear pulses which propagate without change of shape but which also retain their identity subsequent to a localized interaction. In some instances it is possible to find solutions which, while retaining their identity as they propagate, lose it on interacting with other pulses. Such solutions are strictly not solitons but solitary pulses. An example is a Gunn domain: an electric pulse which propagates without change of shape through certain semiconductors. In this book, unless it is necessary to be more specific, all non-linear pulses will be called solitons.

6.2 DISPERSION

To study the basic properties of solitons it is first necessary to discuss the phenomenon of dispersion. Dispersion is a linear phenomenon and hence, to study the dispersive properties of any disturbance, we can express this disturbance as a sum (or integral) of all the individual modes that can propagate in a medium. Thus a disturbance P in a spatially one-dimensional medium can be expressed as

$$P(x,t) = \int dw \int dk \, A(w,k)e^{ikx - iwt} \ . \tag{6.1}$$

Here $A(w,k)$ is the amplitude of the particular mode with frequency w and wave number k. Since each mode can, on its own, propagate in the medium, w and k must be related and satisfy a relation which we may write in the form

$$D(w,k) = 0 \ . \tag{6.2}$$

This is called a dispersion relation. The whole of the physics of the medium is contained in D. For electromagnetic waves propagating in a vacuum one has $D \equiv w^2 - k^2 c^2$, where c is the speed of light. The dispersion relation has two solutions, $w = \pm kc$, corresponding to waves propagating in the positive and negative x direction. The dispersion relation for sound waves takes the same form but now c is the speed of sound. For waves in a warm plasma $D = w^2 - w_p^2 - k^2 V_T^2$ where w_p is the plasma frequency and V_T the thermal velocity. An example of how a dispersion relation is derived is given in Appendix A3.

Since w must satisfy the dispersion relation (6.2), the double integral in (6.1) collapses to a single one and P takes the form

$$P(x,t) = \sum_j \int dk \, B_j(k)e^{ikx - iw_j(k)t} \ , \tag{6.3}$$

where $w_j(k)$ denotes the distinct roots of (6.2), and the summation is over the total number of these roots. Now P, being a physical quantity, must be real. Taking the complex conjugate of (6.3), remembering that P is real, and changing k to $-k$ in the integrand gives (* denotes complex conjugate),

$$P(x,t) = \sum_j \int dk \, B_j^*(-k)e^{ikx + iw_j^*(-k)t} \ , \tag{6.4}$$

For propagating waves which are stable and undamped, the w_js are real, in which case the identification of the right-hand side of (6.3) with that of (6.4) gives either $w_j(k) = - w_j(- k)$ or $w_j(k) = - w_l(- k)$ with $l \neq j$. The first situation applies to sound and electromagnetic waves whilst the second applies to plasma waves where $w_1(k) = + \sqrt{w_p^2 + k^2 V_T^2}$ and $w_2(k) = - \sqrt{w_p^2 + k^2 V_T^2}$.

Now consider more general media but still only those which allow sound-like propagation for waves with large enough wavelength. Then the solution of (6.2), consistent with the condition $w(k) = - w(- k)$, is

$$w(k) = kc - \alpha k^3 + O(k^5) , \tag{6.5}$$

where c and α are real quantities. In the following the terms $O(k^5)$ are ignored.

The phase velocity, defined such that $V_p = w/k = c - \alpha k^2$, is, for a non-zero value of α, a function of the wave number k. Media in which V_p is a function of k are called dispersive. To clarify the meaning of this term consider the form of $P(x,t)$ when only one type of wave is propagating, that is, only one value of j in (6.3), and furthermore, that at $t = 0$, $P(x,t = 0) = \delta(x)$. Then from the properties of the delta function, $B(k) = 1/2\pi$, so that for all x and t

$$P(x,t) = \frac{1}{2\pi} \int_{-\infty}^{+\infty} dk \ e^{ik(x - ct) + i\alpha k^3 t} .$$

This integral can be evaluated and expressed in terms of Airy's integral but in the present context it is sufficient to consider the various spatial moments:

$$\bar{x}(t) = \int_{-\infty}^{+\infty} x \ P(x,t) dx$$

$$= \int_{-\infty}^{+\infty} e^{-ikct + i\alpha k^3 t} \frac{dk}{2\pi} \int_{-\infty}^{+\infty} x \ e^{ikx} dx$$

$$= \int_{-\infty}^{+\infty} e^{-ikct + i\alpha k^3 t} \frac{d}{dk} \delta(k) \ dk$$

$$= ct .$$

Thus the centre of mass, or the average position of the pulse, moves with the speed c. The shape of the pulse is described by higher moments which may be obtained in a similar manner. Thus

$$\overline{(x - \bar{x})^2} = 0 ,$$

whilst

$$\overline{(x - \bar{x})^3} = 6\alpha t .$$

The fact that this last moment is non-zero shows that as t increases the pulse broadens (but in an asymmetric manner), that is, the pulse disperses, and the fact that this broadening simply depends on α being non-zero, shows that dispersion is a universal phenomenon. Only in special circumstances, such as electromagnetic waves in a vacuum, does one have $\alpha \equiv 0$, in which case there is no dispersion.

It is implied by the above discussion that (6.5) describes the w,k relation for all long-wavelength disturbances, at least in one spatial dimension. Thus, although the equations describing particular problems are all different (different dependent variables, different forms), the solution of the corresponding dispersion relation (6.2) reduces to the form (6.5) for long-wavelength disturbances.

Now consider the equation

$$\frac{\partial u}{\partial t} + c\frac{\partial u}{\partial x} + \alpha\frac{\partial^3 u}{\partial x^3} = 0 \ . \tag{6.6}$$

This is an equation with constant coefficients and hence one may assume a solution of the form

$$u = A\mathrm{e}^{ikx - iwt} \ ,$$

where A is some constant. Substitution of this form into (6.6) gives

$$iA(-w + kc - \alpha k^3)\mathrm{e}^{ikx - iwt} = 0 \ ,$$

and hence, for a non-trivial solution to exist, one must have

$$w = kc - \alpha k^3 \ ,$$

or

$$D \equiv w - kc + \alpha k^3 = 0 \ .$$

Thus equation (6.6) captures all the properties of small-amplitude disturbances propagating in a dispersive medium (in the long-wavelength limit). For application to a particular problem, u, c and α must be specified. Note that u equals a constant is a solution of (6.6) and this may be considered to be the equilibrium state.

6.3   THE KdV EQUATION

A question that may be asked is whether it is possible to extend (6.6) to include non-linear effects and still retain its universality. The answer is 'yes' and the KdV equation does just this. The desire (i) for simplicity, (ii) that $u =$ constant remains a solution and (iii) that the non-linearity is not specific, suggests the following equation

$$\frac{\partial u}{\partial t} + \beta\frac{\partial u^2}{\partial x} + \alpha\frac{\partial^3 u}{\partial x^3} = 0 \ . \tag{6.7}$$

The non-linearity and dispersion have been kept separate and $u = u_0$, a constant, is still a solution. Writing $u(x,t) = u_0 + \delta u(x,t)$, substituting into (6.7) and linearizing (that is, neglecting terms such as $\delta^2 u$) shows that δu satisfies (6.6) with $c \equiv 2\beta u_0$. Thus (6.7) correctly includes the relevant dispersive effects. The form of the non-linearity is at this stage much more problematic. Note, however, that if in the term proportional to c in (6.6), u had been replaced by $u + \beta u^2/c$, then a change of coordinates $y = x - ct$ (to a moving frame) reduces the equation to (6.7), but now with x replaced by y.

The above, of course, does not constitute a derivation. For a particular example, namely ion-acoustic waves in a plasma, a formal derivation of equation (6.7) is given in Appendix A3. Similiar derivations have been made for many different types of waves in many different media. The net result is (6.7) with the different physics entering through different values of α and β and meaning of u. All these derivations have one thing in common: the non-linear effects are treated as weak, so that an arbitrary non-linear function can be expressed as a Taylor series in the amplitude u. The non-linearity and the dispersion are then considered as equally important and so (6.7) naturally results. However, there are important cases where, owing to symmetry considerations, for example, the relevant coefficient in the Taylor series is identically zero. The non-linear term u^2 in (6.7) is then replaced by u^3. There are so many cases where (6.7) has been shown to be the relevant equation that it has a high degree of universality. It should be remembered that (6.7) strictly applies to weakly non-linear long-wavelength disturbances, although its range of validity is far from being understood.

By a simple change of variable ($u \rightarrow -6\alpha u/\beta$, $t \rightarrow t/\alpha$) equation (6.7) may be written in the standard form

$$\frac{\partial u}{\partial t} - 6u\frac{\partial u}{\partial x} + \frac{\partial^3 u}{\partial x^3} = 0 \ . \tag{6.8}$$

This is the KdV equation. Notice that by making this change of variables all the physics, that is, the values of α and β, has been subsumed into the scaling.

6.4   PROPERTIES OF THE KdV EQUATION

To illustrate the balance between non-linearity and dispersion it is convenient to look for travelling wave solutions, that is, solutions of the equation which are a function of $y = x - vt$ only. Here v is a constant. The KdV equation then reduces to

$$- (v + 6u)\frac{du}{dy} + \frac{d^3 u}{dy^3} = 0 \ ;$$

with the substitution $u = \bar{u} + v/6$ this equation may be integrated twice to give

$$\frac{1}{2}\left(\frac{d\bar{u}}{dy}\right)^2 = b + a\bar{u} + \bar{u}^3 \ , \tag{6.9}$$

where a and b are constants of integration. The particular solution of interest is the one which goes to a constant and whose derivatives go to zero as $y \rightarrow \pm \infty$. Denoting this constant by $\bar{u}_0$, then $a = -3\bar{u}_0^2$, $b = 2\bar{u}_0^3$ and

$$\frac{1}{2}\left(\frac{d\bar{u}}{dy}\right)^2 = (\bar{u} - \bar{u}_0)^2 \, (\bar{u} + 2\bar{u}_0)$$

so that

$$\int^{\bar{u}} \frac{d\bar{u}}{(\bar{u} - \bar{u}_0)\sqrt{\bar{u} + 2\bar{u}_0}} = \sqrt{2}y$$

The integral is standard and when evaluated gives

$$\bar{u} = \bar{u}_0 - 3\bar{u}_0 \operatorname{sech}^2 \left(\sqrt{\left(\frac{3\bar{u}_0}{2}\right)}, y \right),$$

so that

$$u = \bar{u}_0 - v/6 - 3\bar{u}_0 \operatorname{sech}^2 \left(\sqrt{\left(\frac{3\bar{u}_0}{2}\right)} (x - vt) \right),$$

The particular solution which goes to zero as $x \to \pm \infty$ is obtained for $\bar{u}_0 = v/6$ and then

$$u = - (v/2) \operatorname{sech}^2 \left(\frac{\sqrt{v}}{2} (x - vt) \right). \tag{6.10}$$

Notice that since v must be positive, to keep u real, these solutions propagate in the positive x direction only.

This solution is of the form of a moving pulse, a soliton, which moves with constant speed v. Its amplitude is proportional to v whilst its width is inversely proportional to $\sqrt{v}$. This relationship between amplitude, speed and width is characteristic of a KdV soliton, and is used in the analysis of experimental results to identify a soliton-type solution.

Besides the pulse-like solution, equation (6.9) has an infinite number of periodic solutions. If the conditions on a and b imposed above are relaxed, these solutions can be studied using phase plane analysis as described in Chapter 4. The phase plane portrait for (6.9) is sketched in Fig. 6.1 which clearly shows the existence of periodic

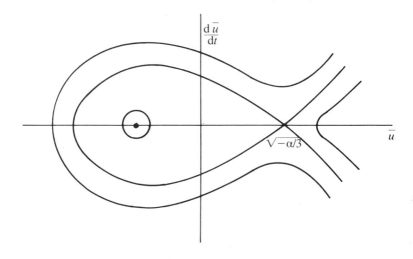

Fig. 6.1 — Phase plane portrait corresponding to the KdV equation (6.9) for $a<0$. For $a>0$ no bounded solutions exist.

solutions corresponding to closed curves about a centre. The soliton corresponds to the trajectory that starts and finishes at the saddle point. Such an orbit trajectory is called a homoclinic orbit. (A trajectory which starts at one saddle point and finishes at another is called a heteroclinic orbit.) The periodic solutions can also be expressed in closed analytic form in terms of elliptic functions and were given many years ago by Korteweg and de Vries. However, because of their phenomenal stability properties, it is the solitons which play the major role in practice.

To obtain some understanding of this implied stability we consider the possible existence of constants of motion. In the following we shall restrict attention to solutions of the KdV equation which vanish exponentially as $x \to \pm \infty$. Then integration of (6.8) over all x gives

$$\frac{d}{dt} \langle u \rangle = 0$$

where

$$\langle u \rangle \equiv \int_{-\infty}^{+\infty} u(x,t)dx \ .$$

Thus we say that $\langle u \rangle$ is a constant of the motion. (Note that we are no longer assuming solutions of (6.8) which are functions of $y = x - vt$ only.) Similarly it is easily shown, using integration by parts, that $\langle u^2 \rangle$ is also a constant of motion. This is not too surprising, as $\langle u \rangle$ and $\langle u^2 \rangle$ are related to momentum and energy. What is surprising is that there are an infinite number of them.

To study this aspect consider the extended KdV equation

$$P \ W \equiv \frac{\partial W}{\partial t} - 6(W + \eta^2 W^2) \ \frac{\partial W}{\partial x} + \frac{\partial^3 W}{\partial x^3} = 0 \ . \tag{6.11}$$

For $\eta \equiv 0$ this reduces to the KdV equations, whilst with the scaling $W = \overline{W}/\eta^2$, $x = \bar{x}\eta^2$ and $t = \overline{t}\,\eta^6$ and in the limit $\eta \to 0$ it reduces to the so-called modified KdV equation

$$\frac{\partial \overline{W}}{\partial t} - 6\overline{W}^2 \ \frac{\partial \overline{W}}{\partial x} + \frac{\partial^3 \overline{W}}{\partial \bar{x}^3} = 0 \ . \tag{6.12}$$

If in the KdV equation (6.8), we make the substitution

$$u = W + \eta \ \frac{\partial W}{\partial x} + \eta^2 \ W^2 \ , \tag{6.13}$$

the equation separates into the form

$$\left(1 + q \ \frac{\partial}{\partial x} + 2\eta^2 W\right) PW = 0 \ . \tag{6.14}$$

Thus any solution of (6.11) is such that u, as defined by (6.13), satisfies the KdV equation. (Note that this does not imply that all solutions of the KdV equation are solutions of (6.11) for the simple reason that there are solutions to (6.14) other than $PW = 0$.)

This is an example of an interesting concept. The solution of distinct non-linear partial differential equations may be linked by simple non-linear transformations such as (6.13).

The relationship between the solutions of the KdV equation and (6.11) is important in respect to the range of validity of the KdV equation. In view of the discussion earlier in this chapter, it is natural to consider (6.11) as the result of taking an extra term in the Taylor series expansion of the non-linearity. This suggests that the solution of a real problem and the solution of the KdV equation may be related in some simple way, such as (6.13), in which case the range of applicability of the KdV equation is greatly enlarged.

Returning to the problem of finding constants of the motion, we expand W as an infinite series in powers of η:

$$W = W_0 + \eta\, W_1 + \eta^2 W_2 + \dots.$$

Substituting this form into (6.13) and equation powers of η (note that u is a given function and is hence independent of η) gives $W_0 = u$, $W_1 = -\partial u/\partial x$, $W_2 = \partial^2 u/\partial x^2 - u^2$, whilst the other coefficients are obtained by solving simple algebraic equations (not differential ones). Now since (6.11) can be written in the form of a conservation equation

$$\frac{\partial W}{\partial t} = \frac{\partial}{\partial x}\left\{3W^2 + 2\eta^2\ W^3 - \frac{\partial^2 W}{\partial x^2}\right\} = 0 \ ,$$

simple integration over all x shows that

$$\frac{\mathrm{d}}{\mathrm{d}t}\langle W_n\rangle = 0 \ ,$$

for all n. Thus we have generated, using purely algebraic techniques, an infinite set of constants of motion for the KdV equation.

Furthermore by the method of construction they are independent. The W_ns give $\langle u\rangle$, $\langle u^2\rangle$ and $2\langle u^3\rangle + \langle(\partial u/\partial x)^2\rangle$ as appropriate constants, the first two agreeing with those obtained by simple integration.

The existence of an infinite set of constants of motion raises an interesting possibility. In classical mechanics it is known that if a sufficient number of constants of motion can be found, the problem of finding the solution is reduced to simple quadrature. Such systems are called integrable. For example, in the case of the simple pendulum, as modelled by the second-order differential equation (4.1), the energy is a constant. This is expressed mathematically by (4.3), where C is essentially the energy. This equation is easily solved for $\mathrm{d}\theta/\mathrm{d}t$ so that

$$\int^{\theta} \frac{\mathrm{d}\theta}{\sqrt{C + w^2\ \cos\theta}} = \sqrt{2t} \ .$$

The problem has been reduced to the evaluation of an integral, that is, to quadrature. A dynamical system with n degrees of freedom requires knowledge of $n-1$ constants of motion for this procedure to work. Now the KdV equation is a partial differential equation, which can, for example by expressing the solution as a Fourier series, be reduced to an infinite set of coupled equations. This infinite set, as seen above, has an infinite number of constants of motion suggesting, but not proving, that the solution of the KdV equation should be expressible in closed analytic form. This in fact turns out to be the case but before going on to consider this aspect in more detail it should be mentioned that other equations share the property of having an infinite number of constants of motion and hence of being integrable. Two such equations, which also have a wide range of physical applicability, are the sine–Gordon equation

$$\frac{\partial^2 \phi}{\partial t^2} - \frac{\partial^2 \phi}{\partial x^2} = \sin \phi \ .$$

and the non-linear Schrödinger equation

$$i \frac{\partial \phi}{\partial t} + \frac{\partial^2 \phi}{\partial x^2} = -|\phi|^2 \phi \ .$$

Equations having this integrability property are in the vast minority but fortunately seem to be the ones that are physically relevant. A number of methods now exist that may be used to obtain solutions of integrable equations in closed form. Here we choose the so-called inverse scattering method to illustrate how to obtain a class of solutions of the KdV equation.

6.5   THE INVERSE SCATTERING METHOD

The problem is to solve (6.8) subject to the initial condition that $u(x,t=0) = u_0(x)$. The method of solution is not straightforward and is in fact a good example of lateral thinking by its originators. The problem is broken down into a number of distinct steps each of which is easily solved or whose solution is known from previous studies.

For a moment forget this problem and consider the linear ordinary differential equation

$$\frac{d^2 \Psi}{dx^2} + (E - u(x,t))\Psi = 0 \ , \tag{6.15}$$

where u is the solution of the KdV equation. The above is nothing but the time-independent Schrödinger equation, familiar from quantum mechanics, with $u(x,t)$ the potential. Time t only enters this equation as a parameter and thus both Ψ and E may be functions of t. Of course the present analysis has nothing to do with quantum mechanics, but the language of this discipline is used to describe the solutions and the method of solution.

At $t=0$, $u(x,t=0) = u_0(x)$, a known function, and in principle one could solve (6.15) for E and ψ. There are two types of solutions: (a) bound states for which $\Psi \to 0$ as $x \to \pm \infty$ and (b) continuum states for which $\Psi \to \exp(ikx)$ in this limit. For simplicity we consider that $u_0(x)$ is chosen such that only bound states exist. From

(6.15), $u(x,t) = (d^2\Psi/dx^2 + E\Psi)/\Psi$, and if this is substituted into the KdV equation, (6.8), we find

$$\frac{dE}{dt}\,\Psi^2 = \frac{\partial}{\partial x}\left\{\frac{\partial \Psi}{\partial x}\,Q - \Psi\,\frac{\partial Q}{\partial x}\right\}\,, \tag{6.16}$$

where

$$Q = \frac{\partial \Psi}{\partial t} + \frac{\partial^3 \Psi}{\partial x^3} - 3(u + E)\,\frac{\partial \Psi}{\partial x}\,. \tag{6.17}$$

Integration over all space, gives $dE/dt = 0$ (Step 1) (since we are restricting attention to bound states for which $\Psi \to 0$ as $x \to \pm\infty$.) Then from (6.15), in the limit as $x \to \pm\infty$ (since we are only considering solutions of the KdV equation with $u \to 0$ as $x \to \pm\infty$)

$$\frac{d^2\Psi_a}{dx^2} + E\,\Psi_a = 0\,,$$

where the subscript a denotes an asymptotic value. Thus we have

$$\Psi_a(x,t) = C_n(t)^{-k_n x}\,. \tag{6.18}$$

where $E_n = -k_n^2$, n labels the possible distinct states and C_n their amplitude. Using the result that $dE/dt = 0$ and the asymptotic nature of Ψ, it is easily shown that the only way that (6.16) can be satisfied is that $Q = 0$. Considering the asymptotic form of $Q = 0$, together with the solution (6.18), leads to the equation

$$\frac{dC_n}{dt} = 4k_n^3 C_n\,,$$

so that (Step 2)

$$\Psi_a(x,t) = \bar{C}_n e^{4k_n^3 t - k_n x}\,.$$

The constants $\bar{C}_n$ may be obtained by solving (6.15), in the asymptotic region, at $t = 0$, since $u(x,t = 0)$ is specified by $u_0(x)$. Then the asymptotic value of Ψ is known for all t.

A similar procedure may be carried out for the continuum states, if they exist. It is found that

$$\Psi_a = e^{-ikx} + R(k,t)e^{ikx}\,,$$

for $x \to \pm\infty$, with

$$R(k,t) = \bar{R}(k)e^{i8k^3 t}\,,$$

and $\bar{R}$ is to be determined from the solution of (6.15) at $t = 0$.

What has been shown in the above is that the asymptotic form of the solution of (6.15) for all time is readily obtainable once it is specified at $t = 0$. The remaining problem is to construct the potential, that is $u(x,t)$, given this asymptotic information. This is identical to the quantum mechanical problem of constructing the potential from the scattering data with $R(k,t)$ simply related to the scattering cross-

section. This inverse problem (hence the name inverse scattering method) was
solved by Gelfand and Levitan who showed that (Step 3)

$$u(x,t) = -2\frac{d}{dx}(K(x,x,t)) ,$$ (6.19)

where K satisfies the integral equation

$$K(x,y,t) - B(x+y,t) + \int_x^\infty dy'\ B(y+y',t)K(x,y',t) = 0 ,$$ (6.20)

and where

$$B(y,t) = \frac{1}{2\pi}\int_{-\infty}^{+\infty} R(k,t)e^{iky}dk + \sum_n C_n^2(t)e^{-k_n y} ,$$

and the sum is over all bound states. Note that x appears in (6.20) simply as a
parameter and not a variable.

This looks more complicated than the original equation (6.8). However, what has
been achieved is that a *non-linear* partial differential equation has been reduced to an
ordinary *linear* integral equation (6.20).

If one makes the restriction that $R(k) \equiv 0$, so that $R(k,t)$ is zero for all time, then
(6.20) can be solved analytically. This condition demands that the initial condition
$u_o(x)$ leads to bound states only. Such potentials $u_0(x)$ are called reflectionless.

An important example is where $u_0(x) = -6\mathrm{sech}^2(x)$. Note that this is not a soliton
in that it is not of the form (6.10). The solution of (6.15), with this value of potential,
gives $k_1 = 1$, $k_2 = 2$. With $R(k,t) \equiv 0$, $B(y,t)$ is the sum of two exponentials, and hence
it is natural to consider the solutions of (6.20) as a simple sum of exponentials. This
proves to be the correct form, and, after a little algebra it may be shown that

$$u(x,t) = -12\frac{\{3 + 4\cosh(2x - 8t) + \cosh(4x - 64t)\}}{\{3\cosh(x - 28t) + \cosh(3x - 36t)\}^2} .$$ (6.21)

We now have a complete solution to the KdV equation, valid for all time and space,
and subject to the initial condition $u(x,t = 0) = u_0(x)$.

Historically the inverse scattering method was the first method to show that the
KdV equation was exactly integrable. However, the above explicit solution seems to
have been given first by Hirota using his so-called direct method. Much more
recently the above solution was expressed in a more useful form by Moloney and
Hodnett, namely

$$u(x,t) = -2\{A_-(\theta_2)\mathrm{sech}^2[\theta_1 + H(\theta_2))/2]$$
$$+ 4A_+(\theta_1)\mathrm{sech}^2[(\theta_2 + H(\theta_1))/2]\},$$ (6.22)

where

$$A_\pm(\theta) = (3 \pm e^\theta)^2/[(1 + e^\theta)(9 + e^\theta)]$$
$$H(\theta) = \ln[(9 + e^\theta)/9(1 + e^\theta)] ,$$

and $\theta_1 = 2x - 8t$ and $\theta_2 = 4x - 64t$. This is now of the form of two pulses, but in general these are not of the form of (6.10). However, consider the solution in a moving frame such that $\theta_1 \simeq 0$ and then for $|t| \to \infty$ we have $\theta_2 \simeq 48t$. In this case the above solution reduces to

$$u(x,t) = -2\mathrm{sech}^2(x - 4(t + a)), \text{ for } t \to -\infty$$
$$= -2\mathrm{sech}^2(x - 4t), \text{ for } t \to +\infty \ ,$$

where $a = -(1/8)\ln 9$. A similar analysis where $\theta_2 \simeq 0$, so that now $\theta_1 \simeq +24t$, gives

$$u(x,t) = -8\mathrm{sech}^2(2x - 32t) \text{ for } t \to -\infty$$
$$= -8\mathrm{sech}^2(2x - 32(t + a/8)) \text{ for } t \to +\infty.$$

All these solutions are of the form of isolated solitons, that is they satisfy (6.10). (The phase factors a and $a/8$ are readily incorporated into the general solution (6.10) by a change of the origin of time.)

We are now in a position to understand the solution of (6.21) and (6.22). For $t \to -\infty$ the solution is in the form of two distinct solitons, one centred about $x = 4(t + a)$, and the other about $x = 16t$. For $t = 0$, $u = -6\,\mathrm{sech}^2 x$ and the solitons have merged into a simple hump. Then as $t \to +\infty$, the two solitons re-emerge *without change*, except for simple phase changes $-a$ and $a/8$. Also note that for $t \to -\infty$ the larger amplitude soliton is behind the other, whilst for $t \to +\infty$, the one of larger amplitude has overtaken the other and now travels in front. The large amplitude soliton thus passes through the other and emerges, after a period of interaction, unchanged, except for a simple phase change. This illustrates the remarkable stability of solitons. They interact in a manner more akin to particles than waves. For obvious reasons the above solution is referred to as a two-soliton solution.

The inverse scattering method has been used to solve other non-linear partial differential equations, for example the sine–Gordon equation and equation (1.6). Also other methods of solution have been found, for example Hirota's direct method, and a method based on Bäcklund transformations. There now exist explicit solutions in the form of N-soliton solutions for a small class of equations. Although this class is small, it does seem to incorporate many of the equations important in applications.

6.6 OTHER UNIVERSAL EQUATIONS

In section 6.3 a heuristic derivation of the KdV equation was given, which was then justified in Appendix A3. In what follows we give heuristic derivations of a number of equations which have been found to have wide application.

Many systems are subject to decay processes which are modelled by diffusive-type behaviour. In the long-wavelength limit, the w,k relation, which must satisfy the condition $w(k) = -w^*(-k)$, must be of the form

$$w = kc - \alpha k^3 - iDk^2 \ .$$

This replaces (6.5). Then, introducing non-linearity, as for the KdV equation, one arrives at the Burger–KdV equation

$$\frac{\partial u}{\partial t} - 6u \frac{\partial u}{\partial x} + \alpha \frac{\partial^3 u}{\partial x^3} = D \frac{\partial^2 u}{\partial x^2} . \tag{6.23}$$

For small k the dispersion (6.5) could equally well be written in the form $w = kc/(1 + \alpha k^2/c)$ or $w(1 + \alpha k^2/c) = kc$. Then, writing down a linear equation analogous to (6.6) and introducing a non-linear term as for the KdV equation gives

$$\frac{\partial u}{\partial t} - 6u \frac{\partial u}{\partial x} - (\alpha/c) \frac{\partial^3 u}{\partial x^2 \partial t} = 0 . \tag{6.24}$$

This equation has the advantage that the phase velocity remains finite for all k.

To obtain a three-dimensional analogue of the KdV equation we consider the particular case of ion-acoustic waves. The dispersion relation is obtained in complete analogy to the one-dimensional analysis given in Appendix A3. For perturbed quantities now proportional to $\exp(-iwt + i\mathbf{k}\cdot\mathbf{x})$ one finds that $|\mathbf{k}|^2$ replaces k^2 and thus from (A3.5)

$$w^2 = w_{\mathrm{P}}^2 |\mathbf{k}|^2 \lambda_D^2/(1 + |\mathbf{k}|^2 \lambda_D^2) . \tag{6.25}$$

We now write $|\mathbf{k}|^2 = k_{\parallel}^2 + k_{\perp}^2$ and choose $k_{\parallel}$ to be the wave number in the direction of propagation of the wave, so that $k_{\perp}$ represents perpendicular variation. In the long-wavelength limit we treat $|\mathbf{k}|^2 \lambda_D^2 \ll 1$, but now introduce the subsidiary expansion that $k_{\perp}^2/k_{\parallel}^2 \ll 1$. Then, to lowest significant order, (6.25) takes the form

$$w/w_{\mathrm{P}} = k_{\parallel}\lambda_D - (k_{\parallel}\lambda_D)^3 + \frac{1}{2} \frac{k_{\perp}^2}{k_{\parallel}} \lambda_D .$$

The equivalent linear differential equation may be written in a number of equivalent ways. We choose the following

$$\frac{\partial u}{\partial t} - w_{\mathrm{P}}\lambda_D \frac{\partial u}{\partial x} + w_{\mathrm{P}}\lambda_D^3 \frac{\partial^3 u}{\partial x^3} + \frac{1}{2} w_{\mathrm{P}} \frac{\partial v}{\partial y} = 0$$

and

$$\frac{\partial v}{\partial x} = \frac{\partial u}{\partial y} ,$$

where x and y denote the parallel and perpendicular directions. The non-linearity is introduced, by analogy with the KdV equation, to give the Kadomtsev–Petviashvili (KP) equation

$$\frac{\partial u}{\partial t} - 6u \frac{\partial u}{\partial x} + \frac{\partial^3 u}{\partial x^3} + a \frac{\partial v}{\partial y} = 0 \tag{6.26}$$

and

$$\frac{\partial v}{\partial x} = \frac{\partial u}{\partial y} .$$

All the above equations apply in the long-wavelength limit (weakly dispersive).

In the opposite extreme of strong dispersion it is possible to obtain another universal equation which now describes the amplitude modulation of a linear wave caused by non-linear effects. An heuristic derivation is as follows. Linear analysis applied to any homogeneous medium shows that perturbations, $\delta\rho$, are of the form $\delta\rho = \phi \exp(ik_0 x - iwt)$ where $w = w(k_0)$ is a solution of the dispersion relation and ϕ some constant. It is also to be expected, by analogy with the simple pendulum, for example, as treated in Appendix A2, that w should be amplitude-dependent in a non-linear theory. Since ϕ is in general complex it is reasonable to expect $w = w(k, |\phi|^2)$. Then for weak non-linear effects and k near the dominant wave number k_0 we may expand and write

$$w = w_0 + w_0' (k - k_0) + \frac{1}{2} w_0'' (k - k_0)^2 + \frac{\partial w_0}{\partial |\phi|^2} |\phi|^2 .$$

where $w_0' = \partial w / \partial k$ evaluated at $k = k_0$. If this form is substituted into the above expression for $\delta\rho$ then all the slow variation may be absorbed into the amplitude ϕ, which is no longer constant, but satisfies the equation

$$i \frac{\partial \phi}{\partial t_1} + iw_0' \frac{\partial \phi}{\partial x_1} + \frac{w_0''}{2} \frac{\partial^2 \phi}{\partial x_1^2} - \frac{\partial w_0}{\partial |\phi|^2} |\phi|^2 \phi = 0 ,$$

where $t_1 = (w - w_0)t$ and $x_1 = (k - k_0)x$. Finally a change of variables from x_1, t_1 to $\xi = x_1 - w_0' t_1$ and t gives

$$i \frac{\partial \phi}{\partial t} + \frac{w_0''}{2} \frac{\partial^2 \phi}{\partial \xi^2} - \frac{\partial w_0}{\partial |\phi|^2} |\phi|^2 \phi = 0 .$$

This is the non-linear Schrödinger equation and is the equation for amplitude modulations of a simple wave. The name is given because of the form of the first two terms but of course in the present context this equation has nothing to do with quantum mechanics.

A derivation, based on a multiple-scale perturbation theory, is given in Appendix A4.

QUESTIONS

1. Derive the dispersion relation appropriate to the partial differential equation

$$\frac{\partial \phi}{\partial t} + a \phi - b \frac{\partial^2 \phi}{\partial x^2} = 0 ,$$

   where a and b are both constant.

2. Use the dispersion relation obtained in question 1 together with equation (6.3) with $B = 1/2\pi$ to show that a pulse-like solution of the equation given in question 1 is of the form

$$P(x,t) = (1/2 \sqrt{\pi bt}) \exp(- at - x^2/4bt)$$

   Use this expression to obtain the time variation of the nth spatial moment of P.

3. Obtain a soliton-like solution of the equation

$$\frac{\partial^2 \phi}{\partial t^2} + a\phi - b\phi^3 - \frac{\partial^2 \phi}{\partial x^2} = 0 \ ,$$

(a and b both positive) such that ϕ and derivatives of ϕ with respect to x go to zero as $x \to \infty$.

4. Obtain soliton-like solutions to the sine–Gordon equation

$$\frac{\partial^2 \phi}{\partial t^2} - \frac{\partial^2 \phi}{\partial x^2} = \sin \phi \ .$$

5. It is readily shown by direct substitution that the equation

$$\frac{d^2 \Psi}{dx^2} + (-1 + 2\mathrm{sech}2x)\, \Psi = 0 \ .$$

has a normalized solution $\Psi = (1/\sqrt{2})\,\mathrm{sech}\,x$. This is the only bounded solution. Use this result together with the method of inverse scattering to obtain the solution of the KdV equation (6.8) appropriate to the initial condition $u(x, t = 0) = -2\,\mathrm{sech}^2(x)$.

6. Obtain a soliton-like solution of equation (6.24) and compare it with the solution of the KdV equation.

7. Find the solution of the non-linear Schrödinger equation (1.6) of the form $\Psi(x, t) = \phi(x, t)\exp(iwt)$ where w and ϕ are both real and $\phi(x, t)$ and its spatial derivatives go to zero as $x \to \pm\infty$.

8. A simple finite difference approximation to the partial differential equation given in question 1 takes the form

$$\frac{d\phi_n}{dt} + a\phi_n - \frac{b}{\Delta^2}(\phi_{n+1} - 2\phi_n + \phi_{n-1}) = 0$$

where we have written $\phi_n(t) = \phi(x = n\Delta, t)$ and Δ is the discrete interval in the x direction. By looking for solutions where ϕ is proportional to $\exp(-iwt + ikn\Delta)$ obtain a dispersion relation between w and k. For considering the limit $\Delta \to 0$ compare with the relation obtained for the continuous case as in question 1.

7

First-order non-linear difference equations

The first-order non-linear difference equation, the so-called logistic equation,

$$x_{n+1} = \lambda x_n \left(1 - x_n\right) , \tag{7.1}$$

where λ is a constant, has a negligible direct application to the physical world. Although it may be a reasonable model for the population dynamics of, say, a particular Japanese butterfly (x_n the number of butterflies in the nth year), the real reason for studying this equation in detail is that it illustrates many of the phenomena found in more realistic models of the physical world. An example is discussed at the end of this chapter.

By choosing a value for λ and an initial value x_0, repeated application of (7.1) will yield the value of x_n for any n. The value of x_0 is mapped into the value x_n. For this reason equations of the form (7.1) are called discrete maps. If x_0 is limited to be between zero and one, and λ between zero and 4, then x_n will remain between zero and one for all n.

Until fairly recently, most physical scientists confronted with the above difference equation would have suggested it was sufficient to consider the related differential equation. This is obtained by assuming that x_n varies continuously and smoothly with n, so that $x_n \equiv x(n\Delta)$, where Δ is a small increment, and $x_{n+1} \equiv x(n\Delta + \Delta)$. If this latter quantity is expanded in a Taylor series, and $n\Delta$ equated to τ, then (7.1) reduces to

$$\frac{\mathrm{d}y}{\mathrm{d}\tau} = \frac{(\lambda - 1)}{\Delta} y - \lambda y^2 , \tag{7.2}$$

where $y = x/\Delta$. This is readily solved to give

$$y(\tau) = aAe^{a\tau}/(1 + \lambda Ae^{a\tau}) ,$$

where $a = (\lambda - 1)/\Delta$ and A is simply related to the initial condition. For $\lambda < 1$, $a < 0$, it is seen that $y(\tau) \to 0$ as $\tau \to +\infty$, whilst for $\lambda > 1$, $a > 0$, $y(\tau) \to a/\lambda$ in this limit. That is, $x_n \to 1 - 1/\lambda$ as $n \to \infty$.

What about the time solution of the difference equation? It is reasonable first to look for the equivalent of time-independent solutions. These are such that $x_{n+1} = x_n$, independent of n, and equal to $\bar{x}$ say, and are the equivalent of fixed points in the theory of differential equations. From (7.1) it is readily seen that the solutions are $\bar{x} = 0$ and $\bar{x} = 1 - 1/\lambda$. The latter is identical to that obtained using the continuous approximation. A numerical solution of (7.1) quickly confirms that for any x_0 the solution will approach this asymptotic value as $n \to \infty$. For these reasons it was more or less assumed that the equivalent differential equation captured the main features of the difference equation. (In fact, as will be shown below, this is correct for $\lambda < 3$.) This led ecologists east of the Atlantic to use differential equations whilst North Americans used difference equations to model their problems. Neither thought there was any fundamental difference.

However, there are huge differences. One way of seeing some of them is by studying the linear stability of the fixed points $\bar{x}$. It should be said that most of the results dicussed in this chapter are readily obtained or confirmed using the minimum of computer facilities. Perhaps the best procedure is to do both, that is follow through the analysis and check the results by direct solution of (7.1).) First consider the differential equation (7.2), write $y = a/\lambda + \delta y(t)$, substitute into (7.2) and linearize to give

$$\frac{d\delta y}{dt} = - a\delta y \ .$$

Thus as long as $\lambda > 1$ $(a > 0)$, the asymptotic state exists and is stable. For the difference equation (7.1), we write $x_n = \bar{x} + \delta x_n$, substitute into (7.1) and linearize, to give

$$\delta x_{n+1} = (2 - \lambda)\delta x_n \ .$$

This is a linear *difference* equation with constant coefficients and can be solved in a manner analogous to that used to solve linear *differential* equations with constant coefficients. Thus we assume a solution of the form $A\mu^n$, substitute into (7.2), and then, for non-trivial solutions to exist, we must have

$$\mu = (2 - \lambda) \ ,$$

so that

$$\delta x_n = (2 - \lambda)^n \delta x_0 \ .$$

The condition for stability, namely $\delta x_n \to 0$ for $n \to \infty$, now implies $|2 - \lambda| < 1$, that is $1 < \lambda < 3$.

Unlike the differential equation the state $\bar{x} = 1 - 1/\lambda$ is unstable for $\lambda > 3$.

What happens for $\lambda > 3$? The easiest thing to do is to use a computer. Take $\lambda = 3.2$, select an arbitrary initial value x_0 $(0 < x_0 < 1)$, and iterate (7.1). Then for the number of iterates sufficiently large (at least 20) x_n alternates between two distinct values. That is, the original state $x = 1 - 1/\lambda$ has bifurcated. This is often referred to as a period-doubling bifurcation. Thus one finds $x_n = x_{n+2} = x_{n+4} \ldots = X$ and $x_{n+1} = x_{n+3} = x_{n+5} \ldots = Y$ with X and Y constants. The new equilibrium or asymptotic state oscillates between X and Y. From (7.1) we must have

$$X = \lambda Y(1 - Y)$$

and

$$Y = \lambda X(1 - X) \ .$$

Elimination of Y for example leads to the quartic equation for X. This is reduced to a quadratic equation by noting that the quartic equation must include the solutions $X = 0$ and $X = 1 - 1/\lambda$. Such a procedure finally leads to the solution

$$X_\pm = \{1 + \lambda \pm \sqrt{(\lambda - 3)(\lambda + 1)}\}/2\lambda \ , \tag{7.3}$$

with $X = X_+$ and $Y = X_-$. These solutions only have meaning when they are real, that is for $\lambda \geqslant 3$.

It is important to note that this new asymptotic solution only exists when the old asymptotic solution, $\overline{X} = 1 - 1/\lambda$, goes unstable. In this sense the whole behaviour is analogous to a phase transformation as discussed in Chapter 3 with respect to the differential equation (3.1).

The stability analysis of this new bifurcated equilibrium proceeds as for the simple equilibrium. One writes $x_n = X + \delta x_n$, $x_{n+1} = Y + \delta x_{n+1}$ and of course $x_{n+2} = X + \delta x_{n+2}$. Substitute these expressions into (7.1) with n taking the values n and $n + 1$, linearize and eliminate δx_{n+1} to give.

$$\delta x_{n+2} = (4 + 2\lambda - \lambda^2)\delta x_n \ .$$

For stability $|4 + 2\lambda - \lambda^2| < 1$, that is, $3 < \lambda < 1 + \sqrt{6}$.

By analogy with the above results we expect that for $\lambda \geqslant 1 + \sqrt{6}$ a further bifurcation occurs leading to an asymptotic state consisting of four distinct values. Unfortunately further analysis, using algebraic methods, becomes fearfully complicated. Numerical methods, on the other hand, give valuable information quite readily.

An initial value x_0 say, and a value for λ, are chosen, and (7.1) iterated a large number of times (for the present purpose a number of iterates of order 200 is sufficient; this is to remove the effect of a particular initial condition). The procedure is continued a further large number of times and the values of x_n, obtained from this later set of iterations, recorded. The whole procedure is repeated for a range of values of λ and the recorded values of x_n are then plotted as a function of λ. For $\lambda < 3$ these x_ns will all have the same value ($= 1 - 1/\lambda$) whilst for $3 < \lambda < 1 + \sqrt{6}$ there will be two values, denoted by X and Y in the above. The number will double at each subsequent bifurcation. The result of such a calculation is shown in Fig. 7.1. This clearly shows the bifurcation sequence as a function of λ. Such a sequence is also referred to as a period-doubling sequence.

If λ_m denotes the value of λ at which the mth bifurcation occurs ($\lambda_1 = 3$, $\lambda_2 = 1 + \sqrt{6}$), it is apparent from Fig. 7.1 that the values of λ_m get closer together as m increases. This suggests that the sequence of bifurcations terminates as $m \rightarrow \infty$ with $\lambda_m \rightarrow \lambda_c$ say. To test this, we consider the function δ_m, defined such that

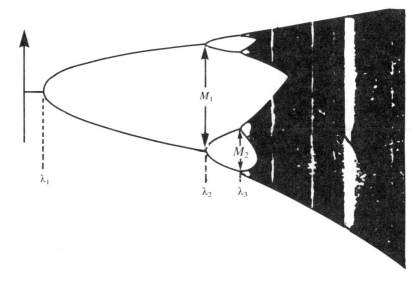

Fig. 7.1 — Bifurcation diagram for the logistic equation (7.1). All possible asymptotic states are shown as a function of the parameter λ.

$$1/\delta_m = (\lambda_{m+1} - \lambda_{\mathrm{m}})/(\lambda_m - \lambda_{m-1}) \ ,$$

as a function of m. It is found, after careful computing, that δ_m approaches the limiting value

$$\delta_{\mathrm{F}} = 4.66920 \ldots \ , \tag{7.4}$$

so that for large m,

$$\lambda_m = \lambda_{\mathrm{c}} - c/\delta_{\mathrm{F}}^m \ , \tag{7.5}$$

where $\lambda_{\mathrm{c}} = 3.568$ and c some constant. The bifurcation diagram may also be characterized by a number M_m, essentially the amplitude of the mth bifurcation. Both λ_m and M_m are shown in Fig. 7.1. The M_ms also form a limiting sequence and it is found that

$$\lim_{m \to \infty} \frac{M_{m+1} - M_m}{M_m - M_{m-1}} = 1/\alpha_{\mathrm{F}} \ ,$$

where

$$\alpha_{\mathrm{F}} = 2.5029 \ldots \ . \tag{7.6}$$

It is interesting that a terminating bifurcation sequence exists, but once this is accepted, the fact that two numbers α_{F} and δ_{F} exist, and describe the sequence, is no great surprise. What is a big surprise, as first found by Feigenbaum, is that terminating bifurcation sequences occur for a subclass of difference equations of the form

$$x_{n+1} = F(x_n, \lambda) \tag{7.7}$$

with *identical* values for the constants that define the sequence, namely δ_F and α_F. These two numbers are called the Feigenbaum numbers and represent a universality of one-dimensional difference equations such as (7.7). Apart from the conditions $F(0,\lambda) = F(1,\lambda) = 0$, which can always be satisfied by a suitable scaling, the maximum of $F(x,\lambda)$, as a function of x, must be sufficiently sharp. The critical value λ_c is not universal but depends on the detailed form of $F(x,\lambda)$. In this way the bifurcation sequence is analogous to a phase transformation. Near the critical temperature T_c the order parameter behaves as $(T_c - T)^\beta$ with β a universal number but T_c model-dependent.

The universality of the bifurcation sequence can be illustrated by the following treatment which is a simple example of renormalization group theory. It is convenient to rewrite (7.1) in the analogous form $y_{n+1} = 1 - ay_n^2$ ($y = (a/\lambda)(x - 1/2)$, $a = \lambda(\lambda/4 - 1/2)$) then

$$y_{n+2} = (1 - a) + 2a^2 y_n^2 - a^3 y_n^4 \ .$$

This, of course, is nothing more than a map for every other iterate of the original map. If we now restrict attention to the iterate that takes a value of x close to $\frac{1}{2}$, that is, around the maximum, then we may neglect the y^4 term in the above equation. This is nothing but a quadratic map and can be readily expressed in the same form as the initial map namely.

$$\overline{y}_{m+1} = 1 - \overline{a}\overline{y}_m^2 \ , \tag{7.8}$$

where we have taken $m = 2n$, $\overline{y}_m = (1 - a)y_{2n}$ and $\overline{a} = 2a^2(a - 1)$. This whole process can now be repeated and, by suitable rescaling (renormalizing), the equation describing a period-4 bifurcation may be cast in the form of the original equation. Thus the whole process of bifurcation is encapsulated by the change of the parameter a, which, going from a bifurcation of order p to one of order $2p$, changes according to

$$a_{p+1} = 2a_p^2 (a_p - 1) \ .$$

This equation has a fixed point ($a_{p+1} = a_p$) which corresponds to the termination of the bifurcation sequence, and gives a critical value $a_c = (1 + \sqrt{3})/2$. For values of a_p near this critical value we may solve the above equation by linearizing. This gives

$$a_p = a_c - \frac{c}{\delta^p} \tag{7.9}$$

where c is a constant and $\delta = 5.72$. With the relationship given above between λ and a this equation is seen to be identical in form to (7.5) with $\lambda_c \simeq 3.52$ and δ an estimate of the first Feigenbaum number δ_F. This is not a particularly good estimate.

However, the procedure can be used to illustrate the universality of δ but not of λ_c. Starting from a general map of the form (7.7), which is scaled to have a maximum at $x = \frac{1}{2}$, it is readily put in the form (7.8) by performing a Taylor series about $x = \frac{1}{2}$ and writing $x = \frac{1}{2} + \overline{y}[F(\frac{1}{2}) - \frac{1}{2}]$ and $\overline{a} = -(F''(\frac{1}{2})/2)(F(\frac{1}{2}) - \frac{1}{2})$. Here $F''(\frac{1}{2})$ denotes the second derivative of $F(x,\lambda)$ with respect to x evaluated at $x = \frac{1}{2}$. The form of a near the critical point will still be given by (7.9) but now the value of λ_c will depend on

$F''(\tfrac{1}{2})$ *and* $F(\tfrac{1}{2})$. However, the value of δ_F appearing in (7.5) will still be δ, reflecting the universality.

This universality has an important repercussion on the analysis of experimental results. Suppose such results have been found to be of the form of a bifurcation sequence specified by the universal numbers α_F and δ_F. Then they can be modelled by an equation of the form (7.7), with x related to, but not necessarily identical to the measurement and λ to the control parameter. (This allows for scaling between x and λ and the experimental results.) However the form for F is not unique and so in the interest of simplicity it is best chosen to be of the form (7.1).

Returning to a discussion of the properties of (7.1), what happens for $\lambda > \lambda_c$? Numerically it is found that x_n does not settle down to any pattern but behaves in a random manner. No matter how many times this equation is iterated it never repeats itself. Such behaviour is called **chaotic**. In many respects it appears to be random but importantly it is totally deterministic, that is, once the initial value x_0 is chosen, the whole sequence of values of x_n is uniquely determined by (7.1). In this way it is fundamentally different from a random sequence and so is called pseudo-random or deterministic-chaotic. Of course it needs more than the introduction of a name to specify what is meant by a chaotic sequence. To this end one introduces the concept of a Liapunov number.

Consider the problem of solving (7.7) subject to two initial values x_0 and $x_0 + \varepsilon$ ($\varepsilon \ll 1$). After one iterate the difference in the two solutions is given by $\Delta x_1 = F(x_0 + \varepsilon, \lambda) - F(x_0, \lambda)$, which for small ε reduces to $\Delta x_1 = F'(x_0)\varepsilon$, where $F'(x_0)$ is $\partial F(x,\lambda)/\partial \lambda$ evaluated at $x = x_0$. Thus after n iterates, for sufficiently small ε, $\Delta x_n = F'(x_0)F'(x_1) \ldots F'(x_{n-1})\varepsilon$, which we may write in the form

$$\Delta x_{n+1} = (-1)^m \Delta x_0 e^{n\mu_n} ,$$

where

$$\mu_n = \frac{1}{n}\sum_{j=0}^{n} \ln |F'(x_j)| .$$

The prefactor $(-1)^m$ is to account for the m values where F' is negative. However, since it is only specifies the sign of Δx, it will be ignored in what follows.

The limiting value of μ_n, as n tends to infinity, is called the Liapunov number. This we will denote by μ, so that

$$\mu = \lim_{n \to \infty} \frac{1}{n}\sum_{j=0}^{n} \ln |F'(x_j)| . \tag{7.10}$$

The existence of a finite value for μ has been confirmed by numerical simulations. These same simulations have also shown that in most cases μ is independent of the initial condition x_0. Of course μ still remains a function of λ. This behaviour is clearly seen in the case of the periodic solutions found for $\lambda < \lambda_c$. For $1 < \lambda < 3$ it was found that the solution of (7.1) evolved to the constant asymptotic state $\bar{x} = 1 - 1/\lambda$, no matter the value of the initial condition. Thus in evaluating μ it is sufficient to take all the x_js the same and equal to $\bar{x}$, in which case (7.10) reduces to $\mu = \ln(|2 - \lambda|)$. Similarly for $3 < \lambda < 1 + \sqrt{6}$, $\mu = (1/2)\ln(|4 + 2\lambda - \lambda^2|)$. It will be noted that in both these cases $\mu \leqslant 0$ for the range of λs for which they are applicable. This means that for sufficiently large n values, $\Delta x_n \to 0$, a result which reflects the stability and hence existence of period solutions. Note, however, that the value of μ calculated for the state $\bar{x} = 1 - 1/\lambda$ goes positive for $\lambda > 3$. That is, when the state goes unstable the Liapunov number as given by $\ln(|2 - \lambda|)$ becomes positive. However, the value of the true Liapunov number is governed by the existence of the bifurcated stable state and is again negative.

Thus to distinguish a chaotic sequence from periodic one of large period, we define a chaotic sequence as one with a positive Liapunov number. Importantly this means that the solution x_n depends *sensitively* on initial conditions, since no matter how small Δx_0, for sufficiently large n the deviation between two solutions, Δx_n, will become significant. A little later in this chapter it will be shown that for $\lambda = 4$, $\mu = \ln 2$, so that for just 33 iterates the deviation between two originally close solutions is magnified by a factor of 10^{10}. Periodic solutions on the other hand are independent of initial conditions.

In view of the pseudo-random nature of the solution of equations such as (7.7) for $\lambda > \lambda_c$ it is natural to try to describe the solution in a statistical sense rather than concentrate on a particular solution. To this end we define $P_n(x)dx$ as the probability of the nth iterate of (7.7) being in the range x to $x + dx$. (The dependence on λ will not be made explicit in the following, though of course $P_n(x)$ depends on this parameter.) Now, since the value x is mapped from two distinct values x_1 and x_2, where $x = F(x_1) = F(x_2)$, we may write a 'conservation' equation

$$P_{n+1}(x)|dx| = P_n(x_1)|dx_1| + P_n(x_2)|dx_2| \ ,$$

an equation which expresses the fact that one does not lose solutions by iterating. From the definition of x_1 we have $dx = F'(x_1)dx_1$, where the prime denotes differentiation, so that the above equation may be written in the form

$$P_{n+1}(x) = \frac{P_n(x_1)}{|F'(x_1)|} + \frac{P_n(x_2)}{|F'(x_2)|} \ . \tag{7.11}$$

This equation is equivalent to the original map as is readily seen by putting $P_0(x) = \delta(x - x_0)$.

The above equation may also be considered as the equation for the ensemble probability, where each member of the ensemble corresponds to a different initial condition x_0. Of particular importance is the asymptotic state (assumed to exist), where P_n is independent of n, $P(x)$ say, which corresponds, in statistical mechanics, to an equilibrium state. It satisfies the equation

$$P(x) = \frac{P(x_1)}{|F'(x_1)|} + \frac{P(x_2)}{|F'(x_2)|} \ . \tag{7.12}$$

Unfortunately, little is known about such equations but what is known is illustrated below. For periodic solutions $P(x)$ can be expressed explicitly in terms of the values x_p which constitute the periodic solution. Thus for (7.1) and $1 < \lambda < 3$ there is one value of x_p, namely $1 - 1/\lambda$, so that

$$P(x) = \delta(x - 1 + 1/\lambda) \ .$$

For $3 < x < 1 + \sqrt{6}$, the values of x_p are given by (7.3) and now

$$P(x) = \tfrac{1}{2}\{\delta(x - X_+) + \delta(x - X_-)\} \ .$$

The factor 1/2 is introduced to satisfy the normalization condition on $P(x)$.

Numerical simulation shows that the solution of (7.1) with $\lambda = 4$ is chaotic. In this case it is convenient to write $x = \sin^2\theta$ and define $P(\theta)$ such that $P(\theta)d\theta = P(x)dx = P(x)2\sin\theta\cos\theta d\theta$. Further, if we identify x_1 with $\sin^2\theta$, then $x_2 = \sin^2(\theta - \pi)$, $x = \sin^2(2\theta)$ and (7.11) reduces to

$$P(2\theta) = (P(\theta) + P(\pi - \theta))/2 \ .$$

By inspection the solution to this equation is simply $P(\theta)$ a constant. Thus we have the non-trivial result that for $\lambda = 4$

$$P(x) = (1/\pi)/\sqrt{x(1 - x)} \ . \tag{7.13}$$

The above change of variable also simplifies the original map as given by (7.1). Writing

$$x_n = \sin^2\theta_n \qquad \text{for } \lambda = 4 \ ,$$

(7.1) takes the equivalent form

$$\theta_{n+1} = 2\theta_n \ , \qquad 0 \leqslant \theta_n \leqslant \pi/2$$

$$= 2(\pi - \theta_n) \ , \qquad \frac{\pi}{2} \leqslant \theta_n \leqslant \pi \ .$$

Because of its 'shape' such a map is called a tent map. Since $|F'(\theta_n)| = 2$ for all θ_n, the corresponding Liapunov number, as defined by (7.10), must be ln2.

A common procedure in statistical mechanics is to replace time averages by ensemble averages. In the present context the analogous procedure is to replace averages over the iterates of the map by an average with respect to the probability function $P(x)$. Thus the definition of the Liapunov number, as given by (7.10), is replaced by

$$\mu = \int_0^1 P(x) \ln(|F'(x)|)dx \ . \tag{7.14}$$

An immediate consequence is that since $P(x)dx = P(\theta)d\theta$, a change of variable and the fact that $|F'(\theta)| = 2$, gives $\mu = \ln2$ for the logistic map for $\lambda = 4$.

For other values of λ no analytic solutions of (7.1) are known to exist. The difficulty in obtaining analytic solutions can be illustrated in the following manner. If we consider a value of x_1 near $\tfrac{1}{2}$, $x_1 = \tfrac{1}{2} + \varepsilon$ say, and substitute into (7.12), assuming $P(\tfrac{1}{2})$ to be finite, then $P(x) \simeq P(\tfrac{1}{2})/|\varepsilon|$ where $x \simeq \lambda/4$. That is, $P(x)$ has a singularity at $x = \lambda/4$ of the form

$$P(x) \simeq \frac{P(\tfrac{1}{2})}{\sqrt{1 - 4x/\lambda}} \ .$$

Using this form for $P(x_1)$ on the right-hand side of (7.12) shows that $P(x)$ has a singularity at $x = (\lambda^2/4)(1 - \lambda/4)$. Thus one expects $P(x)$ to consist of an infinite number of square-root type singularities and therefore it is not surprising that so few analytic solutions exist. Because of this inherent singular structure, new methods of analysis are called for. It may be that these solutions are 'fractal' in which case the methods introduced by Mandelbrot in the study of fractals may be useful.

The structure of $P(x)$ is simplified if $P(\frac{1}{2}) \equiv 0$, which is the case for the periodic solutions, or where the iterate of 1/2 eventually falls on an unstable fixed point. Then $P(x)$ will only have a finite number of discontinuities. For the case $\lambda = 4$ the point $x = 0$ is the unstable fixed point.

To obtain further information about $P(x)$ and μ it is easiest to resort to numerical computation. The evaluation of μ is quite straightforward and is readily obtained as a function of λ. Even this is extremely complicated, as may have been anticipated, knowing that μ_n varies between 0 and ∞ as λ varies between the two values that determine the limits of a particular order of bifurcation. The main features of the variation of μ with λ for the logistic map are illustrated in Fig. 7.2. Some of the fine structure of μ has been removed by smoothing.

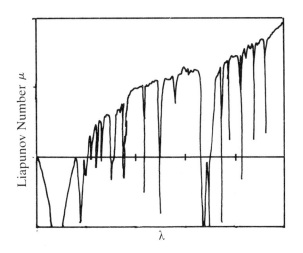

Fig. 7.2 — Variation of the Liapunov number μ, as defined by equation (7.10) or equivalently (7.14), with the parameter λ, for the logistic equation.

The major surprise that emerges from Fig. 7.2 is that in the midst of the chaotic region, $\lambda > \lambda_c$, there are ranges of values of λ where periodic solutions exist, corresponding to the regions where $\mu < 0$. These windows of periodic solutions occupy narrow ranges of λ. The most pronounced is the one associated with a period 3 solution which occurs for $\lambda \simeq 3.8$. Numerical simulations reveal that with change of λ this solution evolves through a bifurcation sequence analogous to that for $\lambda < \lambda_c$ and characterized by the *same* Feigenbaum numbers α_F and δ_F. In fact periodic solutions of any order exist, together with their entourage of bifurcations characterized by α_F and δ_F, but the window of λ values decreases rapidly with the order of the period. All these bifurcation sequences are accompanied by their own region of chaos as is the one for $\lambda < \lambda_c$. All in all, the solution of an equation as simple as (7.1), as a function of both n and λ, can be extremely complicated.

This is one of the reasons why these equations may be useful in modelling experimental situations, such as turbulence, which manifests itself in the form of complicated spatial and/or temporal behaviour. In particular, experiments concerned with the onset of turbulence have been shown to evolve with a change of control parameter, which we can identify with λ, through a sequence of bifurcations before reaching a fully turbulent state. This final state is then identified with a chaotic solution. Because of the experimental difficulty of keeping things constant only a few bifurcations are actually seen. However, analysis of these gives values of α and δ in reasonably good agreement with the Feigenbaum values.

It will be noted from a study of Fig. 7.1 that the period-3 solution emerged from a chaotic solution with an increase in λ. This is in marked contrast with a bifurcation sequence terminating in chaos. The former behaviour is associated with a *tangent bifurcation*, a property as characteristic of difference equations as is a period-doubling bifurcation. To understand this phenomenon consider an equilibrium point of the map (7.7) so that $\bar{x} = F(\bar{x}, \lambda)$. A linear stability analysis about this point leads to the equation.

$$\delta x_{n+1} = F_x(\bar{x}, \lambda)\delta x_n \ , \tag{7.15}$$

where $F_x(\bar{x}, \lambda)$ is $\partial F(x, \lambda)/\partial x$ evaluated at $x = \bar{x}$. The stability criterion is $|F_x(\bar{x}, \lambda)| < 1$. Now denote by x_s and λ_s those values of $\bar{x}$ and λ which make $|F_x| = 1$, the marginally stable point. In the vicinity of this point write $\bar{x} = x_s + \Delta x$ and $\lambda = \lambda_s + \Delta\lambda$, substitute into the equation $x = F(x, \lambda)$, and perform a Taylor series to give to lowest significant order

$$\Delta x = F_x\Delta x + \tfrac{1}{2}F_{xx}(\Delta x)^2 + F_\lambda\Delta\lambda \ ,$$

where $F_\lambda = \partial F(x, \lambda)/\partial\lambda$ and all quantities are evaluated at $x = x_s$ and $\lambda = \lambda_s$. Now $F_x = \pm 1$ and these two cases have to be treated separately.

For $F_x = -1$ the above gives $\Delta x \simeq F_\lambda\Delta\lambda/2$. Then in the vicinity of the marginally stable point, (7.15) reduces to

$$\delta x_{n+1} = \{F_x(x_s, \lambda s) + F_{xx}\Delta x + F_{x\lambda}\Delta\lambda\}\delta x_n \ ,$$
$$= -\{1 - \Delta\lambda(F_{x\lambda} + F_{xx}F_\lambda/2)\}\delta x_n$$

and no matter what the sign of the coefficient of $\Delta\lambda$, a change of sign of $\Delta\lambda$ itself changes a stable state to an unstable state. This is the form of the change which characterizes the bifurcation sequence of $\lambda < \lambda_c$.

Now if $F_x = 1$ the above gives $\Delta^2 x \simeq -2F_\lambda\Delta\lambda/F_{xx}$ and

$$\delta x_{n+1} \simeq (1 \pm \sqrt{-2F_\lambda\Delta\lambda F_{xx}})\,\delta_{xn} \ ,$$

where terms of order $\Delta\lambda$ have been neglected. For one sign of $\Delta\lambda$ the square root is real and then for the negative sign, corresponding to $\Delta x > 0$, $\delta x_n \to 0$ as $n \to \infty$. The state is stable. A change in sign of $\Delta\lambda$ now causes the square root to be imaginary and so $\delta x_n \simeq e^{i\alpha n}$ where $\alpha = \sqrt{|2F_\lambda\Delta\lambda/F_{xx}|}$. The stability of this state is governed by the non-linear effects neglected in the above.

In summary, the behaviour near a stability boundary $(|F_x(\lambda)| = 1)$ depends critically on the sign of $F_x(\lambda)$. For $F_x(\lambda) = -1$ a change of λ will result in a stable state becoming unstable whilst if $F_x(\lambda) = +1$ a stable state changes to one whose stability depends on non-linear effects. Since the conditions $x_s = F(x_s, \lambda_s)$ and $F_x(\lambda_s) = +1$ imply that the line $y = x$ is tangent to the curve $y = F(x, \lambda_s)$ at the point $x = x_s$, the behaviour about this point is called a tangent bifurcation.

It is not possible to satisfy the conditions for a tangent bifurcation directly for the logistic map. However, if we study the map generated by considering every three iterates, that is

$$x_{m+1} = F(F(F(x_m))) \equiv F^{(3)}(x_m, \lambda) \ ,$$

where we have taken $n = 3m$ and replaced x_n by x_m, so that $x_{n+3} \equiv x_{m+1}$, then for $F^{(3)}$ corresponding to $F = \lambda x(1-x)$, it is possible to satisfy the conditions for a tangent bifurcation. This occurs at $\lambda = \lambda_s = 1 + 2\sqrt{2}$ and the fixed point of $F^{(3)}$ corresponds to a period-3 solution of the logistic equation. For $\lambda > \lambda_s$ the period-3 solution is stable, whilst for $\lambda < \lambda_s$ the stability is governed by non-linear terms and numerical simulations show the system to be chaotic.

In short, although the solutions of equations such as (7.1) and (7.7) may be very complicated the general features of the solutions may be understood in terms of period-doubling bifurcations leading to chaos, chaos itself, and tangent bifurcations where periodic solutions emerge from chaotic ones.

Finally as an illustration of how a one-dimensional map can arise from a study of a physical system which evolves continuously with time, we reconsider the Van der Pol equation (4.11) discussed in Chapter 4. There it was shown that the typical solution was of the form of a limit cycle and in the phase plane $(x, dx/dt)$ trajectories spiralled out from the origin, eventually reaching the limit cycle asymptotically. Now, rather than considering the whole trajectory as a continuous function of time, let us consider the value of $x(t)$ at that time when the trajectory crosses the x axis, that is, when $dx/dt = 0$. If we denote such a value of $x(t)$ by x_n then after a certain time interval the trajectory will cross the x axis again. This new value of $x(t)$ we denote by x_{n+1}. Since the Van der Pol equation is a second-order differential equation and we consider only those times where $dx/dt = 0$, then x_n uniquely determines x_{n+1}. Thus an equation of the form (7.7) uniquely specifies the solution of the differential equation (4.11). Unfortunately there seems to be no way of obtaining the form for the map without actually solving the differential equation.

In Chapter 4 an approximate solution of (4.11) was obtained in the form of (4.12). (The appropriate control parameter in this problem is a and this replaces λ in (7.7).) Using this solution it is found that the times for which $dx/dt = 0$ are given by

$$t_n = n\pi/w_0 - \phi/w_0 + (-1)^n(a/2w_0^2)(1 - 7A^2/32) + O(a^2) \ ,$$

which gives, when substituted into the first of (4.12),

$$x_n = (-1)^n A(t_n) + O(a^2) \ .$$

The second equation in (4.12) can be solved to give A as a function of t. Then using the above form for t_n we find to lowest order

$$A^2(t_n) = 4Be^{\alpha n}/(1 + Be^{\alpha n}) \ .$$

where B is a constant of integration and $\alpha = a\pi/w_0$. Defining $y_n = A^2(t_n)/4$, the forms for y_n and y_{n+1} can be used to eliminate B and give a one-dimensional map of the form

$$y_{n+1} = \lambda y_n/(1 + \lambda - 1)y_n) \ ,$$

where $\lambda = e^\alpha$. For small a at least this map is equivalent to the differential equation (4.11).

The above map has an unstable fixed point at $y = 0 (x = 0)$ and a stable one at $y = 1$. This latter fixed point corresponds to the limit cycle solution for $x(t)$.

The map was obtained from the solution of the differential equation by considering the intersection of the trajectories in the phase plane with the line $dx/dt = 0$. This is an example of taking a Poincaré section. The whole procedure is readily generalized. For example, in a three-dimensional phase plane, the Poincaré section could be a surface or even a plane: there is no unique choice, but selection is made so as to simplify the problem. In all cases the result is to replace a differential equation by a difference equation or map. Further examples are discussed in the next chapter.

QUESTIONS

1. Obtain the complete solution of the linear difference equation

$$a\phi_{n+1} + b\phi_n + c\phi_{n-1} = 0$$

   where a, b and c are constants.
2. What is the condition that the equation in question 1 has a solution which is zero for $n = 0$ and $n = N$?
3. Find the simple fixed points of the equation

$$X_{n+1} = a \sin\pi X_n$$

   for $|X_n| < 1$, and use linear analysis to study the nature of the solution in the vicinity of these fixed points.
4. Study the simple fixed points of the symmetric 'tent' map defined such that $X_{n+1} = F(X_n, \lambda)$ where

$$F(X_n, \lambda) = \lambda X_n \text{ for } 0 \leq X_n \leq \tfrac{1}{2}$$
$$= \lambda(1 - X_n) \text{ for } \tfrac{1}{2} \leq X_n \leq 1 \ .$$

5. Obtain the form of the equation for the asymptotic distribution function $P(x)$ for the tent map defined in question 4. Show that for $\lambda = 2$, $P(x) = 1$ is a solution. Use this value to show that the Liapunov number is $\ln 2$.
6. For what value of η does the map $X_{n+1} = F(X_n, \eta)$ with $F(X, \eta) = \eta - 1/X$ have a tangent buifurcation?
7. Show that the map defined in question 6 has an analytic solution of the form

$$X_n = \cos a - \sin a \tan (na + b) \ ,$$

   where a and b are independent of n. How are a and η related?
8. In the study of the Van der Pol equation the following equation was derived

$$y_{n+1} = \lambda \, y_n / (1 + (\lambda - 1)y_n)$$

with $\lambda = e^{\alpha}$. Show that by using a linear change of variable of the form $X_n = s + by_N$ this map may be reduced to the form given in question 6. How does α relate to η? Use the result given in question 7 to obtain an explicit solution for y_n and discuss its asymptotic form as $n \to \infty$.

8

Two-dimensional difference equations

8.1 AREA-PRESERVING MAPS

Like Chapter 7, this chapter starts with a model equation, the so-called standard or Chirikov map,

$$p_{n+1} = p_n - K \sin\theta_n \ ,$$

$$\theta_{n+1} = \theta_n + p_{n+1}, \text{ modulo } 2\pi \ ,$$

(8.1)

where K is a constant. Unlike (7.1) this equation has immediate physical application. It plays an analogous role for difference equations to that of the pendulum equation (4.1) for general oscillatory problems. It was Chirikov who first appreciated the universality of this equation and hence deservedly it is his name which is now attached to it.

This equation is area-preserving. By this one means that a small area in the p, θ phase space, $\Delta p \Delta \theta$, retains its value under iteration, that is, $\Delta p_{n+1} \Delta \theta_{n+1} = \Delta p_n \Delta \theta_n$. Though the area is preserved, the shape is not necessarily the same. It is in fact the changes of shape of an elementary area that lead to the interesting properties of equations such as (8.1).

By definition we may write

$$\Delta p_{n+1} \, \Delta \theta_{n+1} = \begin{vmatrix} \dfrac{\partial p_{n+1}}{\partial p_n} & , & \dfrac{\partial \theta_{n+1}}{\partial p_n} \\[3mm] \dfrac{\partial p_{n+1}}{\partial \theta_n} & , & \dfrac{\partial \theta_{n+1}}{\partial \theta_n} \end{vmatrix} \, \Delta p_n \, \Delta \theta_n .$$

The determinant is called the Jacobian of the transformation. Use of (8.1) shows that the above determinant is unity, and hence (8.1) is such that areas are preserved. It is to ensure area conservation that p_{n+1} appears on the right-hand side of the second of the equations (8.1). The map (8.1) is usually obtained by making approximations to a more fundamental set of equations. If this latter set are area-preserving then it is important that they are modelled by an area-preserving map. Hamiltonian systems are area-preserving, dissipative ones are not. Thus (4.1) is area-preserving while (4.11) is not.

To illustrate how (8.1) can be used to describe a physical situation, consider a ball allowed to fall freely under gravity and bounce on a floor. On contact with the floor the ball is assumed to lose no energy (elastic) and so its velocity is reversed on impact. The equation of motion of the ball while in flight is

$$\frac{d^2x}{dt^2} = -g \ , \tag{8.2}$$

where x is the distance from the floor and g the gravitational acceleration. This is easily solved to give x, as a function of t, in terms of its initial position and velocity v. However, rather than consider the complete trajectory, consider the maximum height h and the time it takes to reach that height. From knowledge of the trajectory it is readily found that after n bounces the height $h_n = v^2/2g$ and the time $t_n = 2nv/g$. Thus the bouncing ball can be described by the difference equations

$$h_{n+1} = h_n$$
$$t_{n+1} = t_n + 2\sqrt{2h_n/g} \ . \tag{8.3}$$

Now consider the case where the height, H, of the floor oscillates periodically with time, such that $H = A\cos(wt)$. The ball bounces elastically with the floor, so on impact its relative velocity is reversed. The exact solution of the trajectory satisfying (8.2) and the boundary condition on impact with the floor can be calculated and the following map obtained

$$h_{n+1} = h_n + wA \ (T_n - t_n)\sin(wT_n) + (A^2w^2/2g)\sin^2(wT_n) \ ,$$

and

$$t_{n+1} = t_n + 2(T_n - t_n) + (Aw/g)\sin(wT_n) \ , \tag{8.4}$$

where T_n is the time the ball hits the floor for the nth time and is given implicitly by

$$A\cos(wT_n) = h_n - g(T_n - t_n)^2/2 \ . \tag{8.5}$$

Of course for $A \equiv 0$ these equations reduce to (8.3). For A small, the first of (8.4) may be written in the form

$$h_{n+1} = h_n + Aw \sqrt{\frac{2h_n}{g}} \sin(wT_n) + O(A^2) \ ,$$

whilst the second of (8.4) and (8.5) may be combined to give

$$T_{n+1} = T_n + \sqrt{\frac{2h_n}{g}} + \sqrt{\frac{2h_{n+1}}{g}} + O(A) \ . \tag{8.6}$$

To order A, h_{n+1} could be replaced by h_n in (8.6) but the above form is retained to ensure that the map is area-preserving. The above equations can now be put in the form of (8.1) by the substitution

$$\theta = wT + \pi, \ p^2 = 8w^2 h/g \text{ and } K = 2Aw^2/g \ .$$

A point that is worth noting is that even for this simple problem, the map (8.1) is not an exact representation. Thus results obtained using (8.1), which depend sensitively on the value of K, should be applied with a certain scepticism.

8.2   THE CHIRIKOV MAP

In Chapter 7 we associated the logistic map (7.1) with a differential equation (7.2). Applying the same procedure to (8.1) gives ($\tau = n\Delta$)

$$\frac{dp}{d\tau} = -(K/\Delta) \sin\theta$$

and

$$\frac{d\theta}{d\tau} = p/\Delta \ ,$$

so that

$$\frac{d^2\theta}{d\tau^2} + (K/\Delta^2) \sin\theta = 0 \ . \tag{8.7}$$

This is an equation describing a pendulum (4.1) whose solution is illustrated in Fig. 4.1(b). Thus for sufficiently small values of K we expect the solution of (8.1) to be of the form illustrated in this figure. This indeed is the case as can be seen by comparison of Fig. 4.1(b) and Fig. 8.1(a). The latter figure was obtained by choosing $K = 0.3$ and iterating (8.1) for a number of different initial conditions. Each contour corresponds to a different initial condition.

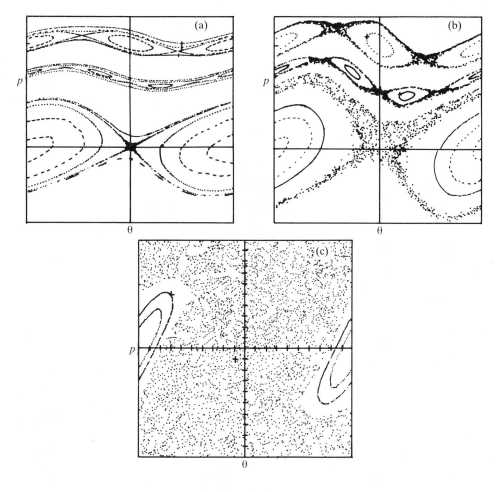

Fig. 8.1 — Phase plane portraits corresponding to the Chirikov map, equation (8.1), for (a) $K = 0.3$, (b) $K = 0.9$ and (c) $K = 4$. Note that in (c) all the points outside the hole arise from just one pair of initial conditions.

For $K = 0.9$ the phase portrait, as shown in Fig. 8.1(b), changes in a qualitative manner. There are now regions in the phase space where the solution appropriate to a *single* initial condition no longer follows a contour, but fills a finite region. Such a solution is now represented by a series of points covering a finite area. The solution is chaotic but some remnants of the order observed for smaller values of K still remain. The regions of chaos remain bounded and there are still contours (trajectories) which segment the phase space. The chaotic trajectories do not cross these contours.

There is a famous theorem, the KAM theorem (named after the three mathematicians Kolmogorov, Arnold and Moser) which in the present context states that for sufficiently small values of K there exist contours in the phase plane which cannot be crossed by trajectories. The phase plane is segmented by these contours and this, as we shall see, has important physical consequences. The numerical results shown in Fig. 8.1(b) for $K = 0.9$ suggest that for this value of K such KAM contours still exist. In fact it has been shown that the last KAM contour disappears for $K = 0.972$, which to all intents and purposes may be taken to be unity. Thus even for $K < 1$, when chaotic solutions exist, all trajectories remain in bounded regions in the phase space.

For $K > 1$ it is possible for chaotic solution to wander through phase space taking all values of p. This is illustrated in Fig. 8.1(c) where $K = 4$. All the points correspond to just one initial condition. KAM contours still exist but they no longer stretch across the whole of the θ axis. (Each closed KAM contour corresponds to a different initial condition.) The chaotic solution wanders through the phase space avoiding the 'hole' defined by the outermost KAM contour.

Closer inspection of the chaotic region reveals that it is interspersed with smaller and smaller 'holes' each with their own set of KAM contours. Although an increase of K results in these holes getting smaller, they never disappear.

The difference in behaviour for $K \leqslant 1$ and $K \gg 1$ is not only mathematically intriguing but has important physical consequences. For example in the case of the bouncing ball, for A sufficiently small such that $K \leqslant 1$, the maximum height reached by the ball will be given in terms of a KAM contour. However, for a sufficiently large value of A, and hence of K, the ball's height may increase without limit. Energy is transferred from the oscillating floor to the ball. This effect forms the basis of a method of isotope separation. By suitable tuning, different isotopes can be made to correspond to different values of K and can hence have different properties in phase space which can be used to separate them. Of course isotopes do not bounce on the floor. The effect of the oscillating floor is replicated by an oscillating field produced by a laser.

The reason for the complicated behaviour found in the iterates of the simple map is associated with resonances. For the bouncing ball problem there are two basic frequencies: one associated with the oscillating floor, that is w, and the other associated with the motion of the ball under the influence of gravity, that is, $w_b = \pi g/V = \pi\sqrt{g/2h}$.

To illustrate this resonant behaviour we consider the case of small K, and make a formal expansion of (8.1) in powers of K. It is first convenient to combine the two equations in (8.1) into one and write

$$\theta_{n+1} - 2\theta_n + \theta_{n-1} = -K \sin\theta_n \ . \tag{8.8}$$

Then writing

$$\theta_n = \theta_n^{(0)} + K\,\theta_n^{(1)} + K^2\,\theta_n^{(2)} + \dots$$

and substituting in (8.8) gives to lowest order

$$L\,\theta_n^{(0)} \equiv \theta_{n+1}^{(0)} - 2\theta_n^{(0)} + \theta_{n-1}^{(0)} = 0 \ . \tag{8.9}$$

The complete solution is $\theta_n^{(0)} = a + pn$, where a and p are arbitrary constants. Substitution of this solution into the first of the equations of (8.1) gives $p_n^{(0)} = p$. To next order in the expansion one obtains from (8.8) that

$$L\,\theta_n^{(1)} = -\sin\,(a + np) \ .$$

This is an inhomogeneous linear equation whose solution is

$$\theta_n^{(1)} = \sin(a + np)/4\sin^2(p/2) \ , \tag{8.10}$$

with

$$p_n^{(1)} = \tfrac{1}{2}\,\{\sin(a + np) + \cot(p/2)\cos(a + np)\}$$

The perturbative method obviously breaks down $(\theta_n^{(1)}/\theta_n^{(0)} \to \infty)$ when $p = 2m\pi$, m any integer. This condition is easily seen to be connected with a resonance since for the bouncing ball $p^2 \equiv 8w^2h/g \equiv 4\pi^2(w/w_b)^2$ and so $w/w_b = m$.

To proceed further with a perturbative expansion it is necessary to reconsider the form of the solution in the neighbourhood of a resonance. This can be done by extending the multiple-scale perturbation theory, as outlined in Appendix A2, to maps. In the spirit of that approach, the constant of integration, in the present context the quantity a, is allowed to be a slowly varying function of n. Thus we write $\theta_n^{(0)} = a_n + np$, which implies $\theta_{n+1}^{(0)} = a_{n+1} + (n+1)p = a_n + (n+1)p + (a_{n+1} - a_n)$, and consider $a_{n+1} - a_n$ to be of order $\sqrt{K}$ and $(a_{n+1} - a_n) - (a_n - a_{n-1})$ to be of order K. Then for $p = 2\pi m$, (8.9) is replaced by

$$L\,\theta_n^{(1)} = -(K\sin a_n + a_{n+1} - 2a_n + a_{n-1}) \equiv S_n \ . \tag{8.11}$$

The operator L controls the fast variation of $\theta_n^{(1)}$ with n whilst S_n only varies on the slow scale. Thus the solution of (8.11) is $\theta_n^{(1)} = n^2 S_n/2$, showing the divergence associated with the resonance. However, as in the conventional multi-scale theory, this divergence may be avoided by suitable choice of a_n, namely

$$a_{n+1} - 2a_n + a_{n-1} = -K\sin a_n \ . \tag{8.12}$$

This is identical to the original equation, but is now supposedly for the slow variation of a. In fact nothing has been accomplished. The resonance is too strong. However, let us proceed and take p to be away from this resonance so that $|p - 2m\pi| \geqslant O(\sqrt{K})$. In this case (8.10) is now an acceptable solution and one can proceed to next order in the expansion in K. This gives

$$\theta_n^{(2)} = (K^2/32)\sin(2(a + np))/\sin^2(p/2)\sin^2(p) \ , \tag{8.13}$$

The expansion now breaks down for $\sin^2(p) = 0(\sin^2(p/2) \neq 0)$, that is, $p = (2n + 1)\pi$. For this value of p we may apply the multi-scale method and now consider $a_{n+1} - a_n = O(K)$. Then (8.12) is replaced by

$$a_{n+1} + a_{n-1} - 2a_n = -(K^2/8)\sin(2a_n) \tag{8.14}$$

and the solution for θ_n with $p = \pi$ becomes

$$\theta_n = a_n + n\pi + (K/4) \sin (a_n + n\pi) + O(K^3) \ , $$

with $p_n = \theta_n - \theta_{n-1}$ and a_n given by (8.14). The explicit dependence of θ_n and p_n on n may be eliminated by expressing p_n as a function of θ_n. This gives

$$p_n = \pi + (K/2) \sin\theta_n - (K^2/16) \sin2\theta_n + (a_n - a_{n-1})(1 - (K/2)\sin\theta_n) + O(K^3) \ . \tag{8.15}$$

The effect of the resonance is contained in the term proportional to $a_n - a_{n-1}$. If this term is neglected then the p_n, θ_n variation is of the form of a simple curve. The resonant term superimposes on this simple curve the whole of the variation described by (8.14), which, of course, is nothing but a rescaled version of the original map. The main effect is that the simple curve is replaced by a sauage-like structure nipped at points corresponding to saddle points of the map (8.14). Adjacent saddle points are separated by a centre.

These features are well illustrated in Fig. 8.2(a) and (b), where the sausage-like structure is shown appropriate to the resonances $p = \pi$ and $p = 2\pi/3$.

The perturbative analysis can be carried out to include the higher resonances at $p = 2m\pi/N$, in which case the variable Na_n satisfies an equation of the form (8.1) with an effective value of $K_{\text{eff}} \propto K^N$.

Importantly, from (8.7), it is seen that the distance between the heteroclinic orbits passing through the saddle points scales as $\sqrt{K}$, so that the width of the separatrices on the Nth resonance is of order $K^{N/2}$. Since $K < 1$, it is seen that for the higher resonances this effect tends to zero and the corresponding trajectories return to simple curves. However, the sausage-like structure is always present, although not easily discernable for small K.

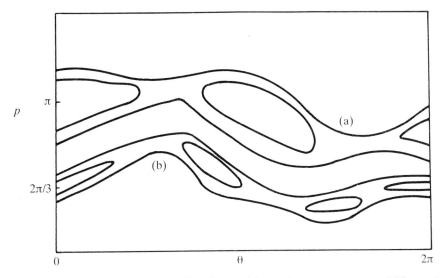

Fig. 8.2 — Analytic construction of KAM curves (a) near the resonance $p = \pi$ and (b) near the
resonance $p = 2\pi/3$.

The motion of a simple pendulum near $\theta = \pi$ (a saddle point) is vulnerable to any
small perturbation. This vulnerability reveals itself in the solution of (8.1), as distinct
from that of (8.7), in that near any saddle point it becomes chaotic. The iterates of the
map no longer lie on curves but fill regions of phase space. This is illustrated in Fig.
8.1 from which it is seen that these chaotic regions increase with increase in K.
Careful analysis reveals that the width of these chaotic regions scale as $\exp(-1/K_{\text{eff}})$
so that the most pronounced region is about the fundamental resonance $p = 0, 2\pi$
where $K_{\text{eff}} = K$.

A major consequence of chaotic regions is that the separatrices no longer
separate regions of the phase space. Orbits, corresponding to iterates of the map, can
now cross the separatrices near the saddle points and move through extended parts of
phase space. This crossing will become more prevalent the larger the value of K.
However, the KAM theorem states that there is a critical value of K, K_c say, below
which there exists at least one line going from $\theta = 0$ to 2π which orbits cannot cross.
Chirikov obtained an estimate for this critical value of K by assuming that the last
KAM line is destroyed when the separatrices of the various resonances overlap. This
gave a value of $K_c \simeq 1$.

Thus for $K > 1$ the chaotic orbits are no longer confined by KAM lines but are
connected across phase space. The case of $K = 4$ is illustrated in Fig. 8.1(c). The
chaotic orbit represented by points corresponds to just one initial condition. Inside
the hole KAM curves still exist with a structure which is a scaled-down version of the
behaviour shown in Fig. 8.1(a).

For larger values of K, the holes in phase space shrink (but never disappear) and
the motion of a typical orbit becomes more and more random.

To quantify this motion we use the first equation of (8.1) to write

$$\langle \Delta p^2 \rangle \equiv \sum_{j=0}^{n} (p_{j+1} - p_j)^2 = K^2 \sum_{j=0}^{n} \sin^2 \theta_j$$

and treat θ_j as a random variable so that $\sin^2 \theta_j$ has an average value equal to $\frac{1}{2}$. Then

$$\langle \Delta p^2 \rangle = K^2 n/2$$

Such behaviour is characteristic of a diffusion process in p with a diffusion constant $D = K^2/4$. More careful calculations show that for $K \geqslant 4$ a diffusion model captures the essential behaviour of the map and the above estimate of D is a reasonable one.

In the context of the bouncing ball problem discussed earlier in this section, the above result means that the height to which the ball can rise increases with the number of bounces n as characterized by a diffusion process. That is

$$h_n = h_0 + \frac{n}{2} (2Aw^2/g) \ ,$$

for $2Aw^2/g \geqslant 4$.

8.3 STRANGE ATTRACTORS (DISSIPATIVE SYSTEMS)

It is a common observation that the introduction of any dissipation in the motion of a pendulum will mean that the pendulum will eventually come to rest ($\theta = d\theta/dt = 0$). The fixed point $\theta = d\theta/dt = 0$, in this case acts as a simple attractor. In Chapter 4 the concept of a limit cycle was introduced. This corresponded to an asymptotic state which is periodic in time and to which neighbouring orbits in phase space are attracted. The limit cycle is an example of a two-dimensional attractor. The generalization to higher-dimensional phase space is apparent: in three dimensions the closed curve forming the limit cycle may be replaced by a toroidal surface or by a closed three- dimensional curve.

These attractors have one thing in common. They all exist in a phase space of dimension higher than their own. Thus the equations of motion describing the possible orbits in the full phase space must be such as to allow this contraction of dimension. Mathematically this is expressed by saying that the Jacobian is less than unity.

In the cases cited above the attractor is simple, that is, the curve or surface is smooth. This was considered to be the most general (generic) type of attractor, although Poincaré had suggested differently. However, in 1971 Ruelle and Takens pointed out that toroidal attractors are structurally unstable and the generic attractor is a strange attractor. The 'strangeness' arises because the simple toroidal surface is replaced by an infinite set of nested toroidal surfaces, somewhat like the skins of an onion. The trajectory, once attracted to this convoluted structure, moves from layer to layer without ever passing through the same point more than once.

The importance of the existence of a strange attractor is that the trajectory or solution is now a complicated non-periodic function of time and thus may be associated with phenomena such as turbulence, which also have a complicated time behaviour. Equally importantly, it has been shown that strange attractors can exist in a three or more dimensional phase space, suggesting that it may be possible to study turbulent-like behaviour by considering a few coupled non-linear first-order differential equations. In fact Lorenz had independently come to this conclusion from a numerical study of equations (1.3). These were derived from the Navier–Stokes equation, a set of partial differential equations, by expanding the spatial variation in Fourier series to obtain an infinite set of coupled ordinary differential equations, and then, somewhat arbitrarily, truncating this set to three. (This procedure is illustrated in Chapter 1 in the derivation of (1.9) from (1.5).)

In two dimensions (two coupled first-order non-linear differential equations) it is not possible to have a strange attractor. The limit cycle is structurally stable and is the most complicated form of attractor that may exist. This no longer holds for difference equations and a particularly simple model of a strange attractor, due to Hénon (1976) is governed by the map

$$X_{n+1} = 1 - aX_n^2 + bY_n \ , \qquad\qquad (8.16)$$

$$Y_{n+1} = X_n \ ,$$

with a and b constant parameters. For $b \equiv 0$ the above reduces to the logistic map, which for $a \geqslant 1.4$ shows chaotic behaviour. The Jacobian for the map is simply b so that $1 - b$ is a measure of the contraction per iterate.

The results of a numerical solution of (8.16) with $a = 1.4$ and $b = 0.3$ is shown in Fig. 8.3. It is found that for all initial conditions contained in a large region about the origin the final state is always the same and is as shown in Fig. 8.3. Thus we have an attractor. The area enclosing all initial points which eventually iterate to this attractor is called the basin of attraction.

What Hénon did was to 'magnify', that is to look more closely at the lines which make up the attractor, and found that a single line on closer examination was really two. Moreover, any of these lines on closer inspection was found to be composed of two and so on and so on. Thus the complexity of the attractor depends on how close one looks. Closer inspection, which means more accurate plotting of the solution, reveals more and more 'skins of the onion'. The complicated transverse structure, corresponding to an infinite set of closely packed lines, gives rise to the strangeness of this attractor. It is important to note that a particular iterate of the map roams chaotically along the many curves of the attractor without ever returning to the same point.

These features of the attractor can be understood by using a simple perturbation theory, treating b as small. To lowest order (neglecting the term proportional to b) one has $X_{n+1} = 1 - aX_n^2, Y_{n+1} = X_n$ or $X_{n+1} = 1 - aY_{n+1}^2$. Since this equation is true for all n it must be the equation for the attractor, so that to lowest order

$$x = 1 - ay^2 \ .$$

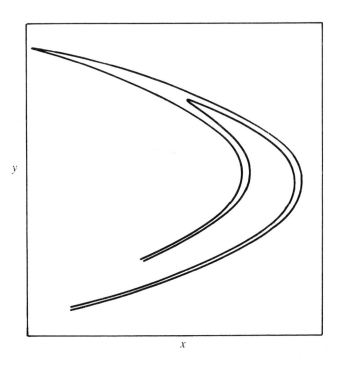

Fig. 8.3 — Phase plane portrait of the Hénon map, equation (8.16), illustrating the complexity of a strange attractor.

To next order we use this result to estimate the term proportional to b and write

$$X_{n+1} = 1 - aX_n^2 \pm \xi \sqrt{1 - X_n} \ ,$$

where $\xi = b/\sqrt{a}$. This is a one-dimensional map but is now double valued. The corresponding equation for the attractor is

$$x = 1 - ay^2 \pm \xi \sqrt{1 - y} \ , \tag{8.17}$$

and we see that what to lowest order was a single curve is now two, separated by a distance proportional to ξ. This expansion procedure can be continued and at each stage a doubling of the number of lines making up the attractor occurs, corresponding to the doubling found numerically by Hénon. Since the procedure can be continued indefinitely the final form for the strange attractor corresponds to an infinite set of nested lines. Such is the complexity of a strange attractor.

　　To next order the equation for the attractor is

$$x = 1 - ay^2 \pm \xi \sqrt{1 - y \pm \xi} \ .$$

However, to proceed further it is necessary to become more formal. The essential nature of the above approximate treatment of a strange attractor is that a two-dimensional map (8.16) has been replaced by a one-dimensional one which, however, is multi-valued (8.17). Thus we write the equation for the strange attractor in the form

$$X_{n+1} = S(X_n) \ , \qquad Y_{n+1} = X_n$$

with $S(x)$ a multi-valued function of x. The above equation may be formally solved for X_n to give $X_n = S^{-1}(X_{n+1})$ so that $Y_n = S^{-1}(X_n)$. If these forms are substituted into (8.16) one gets an equation for the strange attractor of the form $X = S(Y)$, where

$$S(x) = 1 - ax^2 + bS^{-1}(x) \ . \tag{8.18}$$

Unfortunately little is known about how to solve such equations analytically but they do yield to numerical algorithms and have been used to study the more detailed form of the strange attractor. Finally it should be stressed that both the above method and the expansion in ξ only give the shape of the attractor and say nothing about the density of points on the attractor. In fact this density is controlled mainly by the map with $b \equiv 0$. Since in this case it is equivalent to the logistic map with the change of parameter $\lambda = \sqrt{1 + 4a} + 1$, a change of a can give rise to periodic solutions as well as chaotic ones. For some values of a and b the attractor collapses to a few points. These points lie on the curves given by (8.17) or (8.18). This behaviour is analogous to that of the logistic map studied in Chapter 7 where by changing the parameter λ one went from periodic behaviour to chaos.

The question remains how to quantify the transverse complexity of this strange attractor. One way is to calculate the fractal or Handsdorff dimension. This concept is discussed in Appendix A5 and we now apply this general method to calculate the transverse dimension of the attractor. Imagine carrying out the perturbation expansion, outlined above, n times. Then to this order the attractor will consist of 2^n separate lines. These, on average, will be spaced a distance ξ^n apart. In calculating the dimension we take the box size to be $\varepsilon = \xi^n$ in which case the number of filled boxes will be 2^n. Thus, following the procedure given in A5, the transverse dimension $d_\perp$ is defined by

$$d_\perp = \lim_{n \to \infty} \frac{\ln 2^n}{\ln 1/\xi^n} = -\ln 2/\ln \xi \ .$$

This value is confirmed by numerical computations.

The above calculation implicitly assumes the lines in the attractor do not cross.

The effect of crossing, when it takes place, is to put an upper limit on $d_\perp$ of 2 so that the above expression for $d_\perp$ is correct as long as its value is less than 2.

8.4 SUMMARY

In this chapter we have illustrated the most important properties of two-dimensional difference equations. This has been done by choosing two special maps, the Chirikov map (8.1) and the Hénon map (8.16). The Chirikov map has unit Jacobian and is the analogue of a Hamiltonian system whilst the Hénon map has $J = b < 1$ and is dissipative.

In both cases, chaotic solutions can exist for suitable values of the parameters. For non-dissipative systems ($J \equiv 1$), as exemplified by the Chirikov map, the chaos can lead to a diffusion of the solution (trajectory) through phase space. For dissipative systems ($J < 1$), as exemplified by the Hénon map, the chaos reveals itself in that the lines making up the attractor are not continuous but rather a collection of points arranged chaotically along the line.

QUESTIONS

1. Discuss the stability of the simple fixed points of the Chirikov map, equation (8.1). Compare the results with those obtained using the continuous approximation as given by equation (8.7) (see question 1 of Chapter 4).

2. Use the expressions given by (8.10) to show that to $O(K^2)$ the iterates of the map lie on the smooth curve

$$\bar{p}(\theta) = p + (K/2) \{\sin\theta + \cot(p/2)\cos\theta\}$$

3. Show that the particular solution of (8.7) which passes through the heteroclinic points $\theta = \pm\pi$ takes the form

$$\theta = -\pi + 4\tan^{-1}(\exp(\tau\sqrt{K/\Delta^2}))$$

   Use this result to show that the maximum displacement of $d\theta/dt$ is proportional to $\sqrt{K}$.

4. Find the fixed points of the Hénon map, equation (8.16), and examine their stability for small values of b.

5. Find the period-two fixed points of the Hénon map and show such points exist for $a > 3\gamma^2/4$ where $\gamma = 1 - b$.

Appendix A1

Since the Fokker–Planck equation is linear, a formal solution is readily obtained by using the method of separation of variables. Thus one may write

$$P(q,t;q_0,t_0) = \sum_n A_n \, P_n(q) e^{-\lambda_n(t-t_0)} \, , \tag{A1.1}$$

where $P_n(q)$ satisfies the linear equation

$$L \, P_n \equiv D \, \frac{\mathrm{d}^2 P_n}{\mathrm{d}q^2} - \frac{\mathrm{d}}{\mathrm{d}q} \, (H \, P_n) + \lambda_n P_n = 0 \, , \tag{A1.2}$$

subject to the boundary conditions $P_n = 0$, $q = q_1, q_2$. Associated with the linear operator are the adjoint operator L^+ and adjoint functions P_n^+, defined such that

$$L^+ \, P_n^+ \equiv D \, \frac{\mathrm{d}^2 P_n^+}{\mathrm{d}q^2} + H \, \frac{\mathrm{d}P_n^+}{\mathrm{d}q} + \lambda_n \, P_n^+ = 0 \, , \tag{A1.3}$$

and subject to the same boundary conditions as P_n. The adjoint operator is defined such that

$$\int_{q_1}^{q_2} P_n^+ \, L \, P_m \, \mathrm{d}q = \int_{q_1}^{q_2} P_m \, L^+ \, P_n^+ \, \mathrm{d}q \, ,$$

for all m and n. The P_ns and P_m^+s are defined so as to be orthonormal, so that

$$\int_{q_1}^{q_2} P_n(q) \, P_m^+(q)\mathrm{d}q = \delta_{n,m} \; ,$$

with $\delta_{n,m} = 1$, $n = m$ and zero otherwise. In this case one may express the delta function in the form

$$\delta(q - q_0) = \sum_n P_n(q) \, P_n^+(q_0) \; . \qquad (A1.4)$$

Thus, imposing the requirement that $P(q,t_0;q_0 t_0) = \delta(q - q_0)$ on the expression (A1.1) gives the unique solution of the Fokker–Planck equation in the form

$$P(q,t;q_0,t_0) = \sum_n P_n(q) \, P_n^+(q_0)\mathrm{e}^{-\lambda_n(t - t_0)} \; .$$

If this form is substituted into the equation, (3.21), for the mean first-passage time, one gets

$$M(q_0) = \sum_n \frac{P_n^+(q_0)}{\lambda_n} \int_{q_1}^{q_2} P_n(q')\mathrm{d}q' \; .$$

By direct substitution, it is readily found, using (A1.3) that

$$D \, \frac{\mathrm{d}^2 M}{\mathrm{d}q_0^2} + H(q_0) \, \frac{\mathrm{d}M}{\mathrm{d}q_0} = - \sum_n P_n^+ (q_0) \int_{q_1}^{q_2} P_n(q')\mathrm{d}q'$$

and, finally, using (A1.4) that

$$D \, \frac{\mathrm{d}^2 M}{\mathrm{d}q_0^2} + H (q_0) \, \frac{\mathrm{d}M}{\mathrm{d}q_0} = - 1 \qquad (A1.5)$$

with, of course, $M = 0$ for $q_0 = q_1$ and q_2.

Appendix A2:
Multiple scale perturbation theory

Unfortunately most non-linear equations cannot be solved analytically and one must resort to approximate methods. A very powerful method is perturbation theory. Applied to non-linear equations, this method itself introduces an inherent stumbling block. This difficulty, and a way around it, is the subject of this appendix.

Consider the simple pendulum as described by equation (4.1). From the phase diagram, as illustrated in Fig. 4.1(b), it is seen that periodic solutions exist for $|\theta| < \pi$. Now consider applying conventional perturbation theory to this equation. Introduce an expansion parameter ε, expand θ in the form

$$\theta(t) = \varepsilon \theta_0(t) + \varepsilon^3 \, \theta_1(t) + O(\varepsilon^5) \; ,$$

substitute this form into (4.1) and equate powers of ε. To order ε^3 this gives

$$\frac{d^2 \theta_0}{dt^2} + w^2 \theta_0 = 0$$

$$\frac{d^2 \theta_1}{dt^2} + w^2 \theta_1 = w^2 \theta_0^3 / 6 \; .$$

These are readily solved to give

$$\theta_0 = A \, \cos(wt + \phi)$$

and

$$\theta_1 = \frac{A^3}{24} \{ \tfrac{3}{2} \, wt \, \sin(wt + \phi) + \tfrac{1}{8} \cos(3(wt + \phi)) \} \ ,$$

where A and ϕ are constants of integration.

It will be noted that the first term in the expression for θ_1 is *not* periodic no matter how small the value of A. After sufficient time this term will dominate. This is contrary to the form of the exact result, which from a phase plane approach is known to be periodic. It is also contrary to everyday experience. One does not expect a simple pendulum, once set in motion, to oscillate with ever-increasing amplitude.

To give an indication how this basic difficulty may be overcome, one considers the complete solution to this order, namely $\varepsilon\theta_0 + \varepsilon^3\theta_1$, and notes that the troublesome term may be absorbed into θ_0 by allowing a change in phase (or frequency). Thus

$$\theta = \varepsilon \, \theta_0 + \varepsilon^3 \, \theta_1 + O(\varepsilon^5)$$

$$\equiv \varepsilon \, A \cos \left(wt + \phi - \frac{\varepsilon^2}{16} A^2 wt \right) - \frac{\varepsilon^3 A^3}{192} \cos \left(3(wt + \phi) \right) + O(\varepsilon^5) \ . \quad (A2.1)$$

and the new expression, which is identical to the old to $O(\varepsilon^5)$, is now seen to be periodic. One has regained the important qualitative property of periodicity.

The above result may be interpreted in two ways. Firstly, that the non-linearity produces an amplitude-dependent frequency $\bar{w}$ where $\bar{w} = w(1 - \varepsilon^2 A^2/16)$, or secondly, that the phase, now defined as $\phi - \varepsilon^2 w^2 A^2 t/16$, is a slowly varying function of time. It is the second interpretation that it is necessary to invoke in multi-scale (sometimes called multi-time) perturbation theory.

The above procedure could be repeated to higher order, and, at each stage, the troublesome terms removed by a change in phase or possibly a change in the amplitude A. However, this process becomes algebraically complicated and a more systematic approach is needed. This is provided by a multi-scale perturbation theory which has its origins in the work of Poincaré and the method of averaging due to Bogoliubov and Krylov. The method is based on the existence of two distinct time scales: a fast one which in the above example is just $2\pi/w$ and a slow one which is $32\pi/\varepsilon^2 A^2 w$. The time variation on the slow scale is obtained by averaging out the time variation on the fast time scale.

Formally the method proceeds as follows. Instead of the single time t, two times, one $\tau_0 \equiv t$ and the other $\tau_1 \equiv \varepsilon^2 t$, are introduced and treated as *independent* variables. Then writing

$$\theta = \varepsilon \, \boldsymbol{\theta}_0(\tau_0 . \tau_1) + \varepsilon^3 \, \theta_1(\tau_0, \tau_1) + O(\varepsilon^5) \ ,$$

one has, since $\mathrm{d}\tau_1/\mathrm{d}t = \varepsilon^2$,

$$\frac{d\theta}{dt} = \varepsilon \frac{\partial \theta_0}{\partial \tau_0} + \varepsilon^2 \frac{\partial \theta_0}{\partial \tau_1} + \varepsilon^3 \frac{\partial \theta_1}{\partial \tau_0} + O(\varepsilon^5)$$

and

$$\frac{d^2\theta}{dt^2} = \varepsilon \frac{\partial^2 \theta_0}{\partial \tau_0^2} + \varepsilon^3 \frac{\partial^2 \theta_1}{\partial \tau_0^2} + 2 \frac{\partial^2 \theta_0}{\partial \tau_0 \partial \tau_1} + O(\varepsilon^5) .$$

Substitution into (4.1) gives as before

$$\frac{d^2\theta_0}{d\tau_0^2} + w^2 \theta_0 = 0 ,$$

with a solution $\theta_0 = A \cos(w\tau_0 + \phi)$, but now, A and ϕ, which of course must be independent of τ_0, *may be functions of* τ_1. To next order

$$\frac{\partial^2 \theta_1}{\partial \tau_0^2} + w^2 \theta_1 = (w^2 A^3/24)\cos(3(w\tau_0 + \phi))$$

$$+ (w^2 A^3/8)\cos(w\tau_0 + \phi)$$

$$+ 2w \left\{ \frac{\partial A}{\partial \tau_1} \sin(w\tau_0 + \phi) + A\frac{\partial \phi}{\partial \tau_1} \cos(w\tau_0 + \phi) \right\} ,$$

which is an equation in τ_0.

Now the terms which give rise to the non-periodic behaviour of θ_1 are the resonant or secular terms, namely those proportional to the homogeneous solution of the equation. That is, terms proportional to $\sin(w\tau_0 + \phi)$ or $\cos(w\tau_0 + \phi)$. It was the second term on the right-hand side of the above equation which gave rise to the non-periodic behaviour. Now we can use the added freedom of the time variation of A and ϕ to remove all these secular terms. Thus we demand

$$\partial A/\partial \tau_1 = 0$$

and

$$2wA\partial\phi/\partial\tau_1 + w^3 A^3/8 = 0 .$$

Thus A is independent of τ_0 and τ_1 whilst $\phi = \phi_0 - w^2 A^2 \tau_1/16$ where ϕ_0 is independent of τ_0 and τ_1. Combining the above expressions, it is readily seen that the solution (A1.1) is recovered, if, at the end of the calculation, one replaces τ_0 by t and τ_1 by $\varepsilon^2 t$.

To proceed to higher order it is necessary to introduce further time scales $\tau_2 = \varepsilon^4 t$, $\tau_3 = \varepsilon^6 t$ for example. At each stage the variation of quantities such as A and ϕ_0 is used to remove the resonant terms. This ensures a final solution which is periodic.

The expansion parameter ε is introduced to simplify the mathematical book-keeping. The time expansion parameter in the present case is the amplitude A, as is made obvious by the fact that in the final expression for θ only the combination εA appears. Thus at the end of the calculation one simply puts $\varepsilon = 1$.

As stated above conventional perturbation theory breaks down after a sufficiently long time. An estimate of this time can be obtained by demanding $\theta_0 \simeq \theta_1$. Thus for $A^2 wt/16 > 1$ the method is of little use. To obtain an estimate for the two-time analysis given above, it is necessary to consider the resonant terms which appear in the equation for θ_2. The condition $\theta_0 \simeq \theta_2$ leads to the condition $\alpha A^4 wt < 1$ for the range of validity. (α is some number of order 16.) Importantly t now scales as $1/A^4$ compared to $t = 1/A^2$ for conventional theory.

Of course if one introduces higher time scales $\tau_n (= \varepsilon^{2n} t)$ into the analysis then the range of validity is governed by the condition $\bar{\alpha} A^{2(n+1)} wt < 1$.

In the above analysis of the pendulum equation a particular ordering in terms of ε was made in the expansion. *A priori*, this ordering is not known, and is arrived at by intuition and guesswork. It is chosen so as to remove the secular terms, but these are not known until the analysis is partially completed. A rule of thumb is to proceed with ordinary perturbation theory until the difficulties arise, and then introduce an ordering in ε to allow the difficulties to be removed. The method is used to solve a number of distinct problems in this book and consideration of these should help in understanding the procedure of arriving at the correct ordering scheme.

Appendix A3:
Ion acoustic waves and the KdV equation

Consider a plasma composed of ions and electrons. Because of the large difference in their masses, the electrons are enslaved to the ions and hence are in local equilibrium determined by the local electrostatic potential $\phi(x,t)$. Thus the electron density $n_e(x,t)$ is given by

$$n_e(x,t) = N \exp(-e\phi/k_B T) \ . \tag{A3.1}$$

Here T is the temperature (equal for electrons and ions) k_B Boltzmann's constant and $-e$ the electron charge.

The ions are treated as a fluid characterized by a density ρ and velocity v. Conservation of density gives (in one dimension)

$$\frac{\partial \rho}{\partial t} + \frac{\partial}{\partial x}(\rho v) = 0 \ , \tag{A3.2}$$

whilst conservation of momentum gives

$$\frac{\partial v}{\partial t} + v\frac{\partial v}{\partial x} = \frac{e}{m}\frac{\partial \phi}{\partial x} \ . \tag{A3.3}$$

These equations are finally closed by Poisson's equation

$$\frac{\partial^2 \phi}{\partial x^2} = -4\pi e(n_e - \rho) \ . \tag{A3.4}$$

In the above, m is the ion mass.

It is instructive to examine the linear problem. Then ϕ and v may be treated as small quantities whilst $\rho = N + \delta\rho$. The linearized equations then take the form

$$\frac{\partial \delta\rho}{\partial t} + \frac{\partial}{\partial x}(Nv) = 0$$

$$\frac{\partial v}{\partial t} = \frac{e}{m}\frac{\partial\phi}{\partial x}$$

and

$$\frac{\partial^2\phi}{\partial x^2} = 4\pi e\,(\delta\rho + Ne\phi/k_B T)\ .$$

The coefficients of all perturbed quantities in the above equations are constant and so all these quantities are proportional to $\exp(-iwt + ikx)$. Substitution of this form into the above gives rise to a set of algebraic equations which have a non-trivial solution only if the dispersion relation

$$D(w,k) \equiv 1 - k^2\lambda_D^2\left(\frac{w_p^2}{w^2} - 1\right) = 0\ , \tag{A3.5}$$

is satisfied. ($w_p^2 = 4\pi Ne^2/m$, the plasma frequency and $\lambda_D^2 = k_B T/4\pi Ne^2$, the Debye length.)

In the long-wavelength limit ($k \to 0$) the above equation may be solved to give

$$w/w_p \simeq k\lambda_D - \tfrac{1}{2}(k\lambda_D)^3 + O(k^5)\ .$$

Modes of oscillation satisfying this equation are called **ion-acoustic** modes, by analogy with ordinary acoustic (or sound) waves. Then $w_p\lambda_D = \sqrt{k_B T/m}$ is the speed analogous to the sound speed. It will be noticed, though, that unlike sound waves the above waves are dispersive.

It is the purpose of this appendix to obtain, in a certain approximation, the KdV equation, as a replacement for the full non-linear equations given above. The method of approximation is based on the smallness of k and hence of w. Thus the whole analysis is restricted to long-wavelength disturbances. Thus we introduce scaled time and space variables $\tau_1 = \varepsilon t$ and $x_1 = \varepsilon x$. As discussed in Chapter 6, the KdV equation expresses a balance between dispersion and non-linearity. The variables ρ, v and ϕ are expanded in powers of ε so as to reflect this balance. Finally the method of multiple-scale perturbation theory is used to avoid the possibility of secular behaviour. These considerations, together with a little hindsight, suggest the expansion

$$\rho(x,t) = N + \varepsilon^2 \rho_1(x_1,t_1,t_2) + \varepsilon^4 \rho_2(x_1,t_1,t_2) + O(\varepsilon^6),$$

with similar expansions for ϕ and v, though there is no term independent of ε for both these cases. ($t_2 = \varepsilon^3 t$). Substitution of these forms into the basic equations (A3.2), (A3.3) and (A3.4) give to order ε^3

$$\frac{\partial \rho_1}{\partial t_1} + N\frac{\partial v_1}{\partial x_1} = 0$$

$$\frac{\partial v_1}{\partial t_1} - e\frac{\partial \phi_1}{\partial x_1} = 0$$

and

$$\rho_1 + Ne\phi_1/k_B T = 0.$$

These are linear equations with constant coefficients and by assuming solutions proportional to $\exp(-iwt_1 + ikx_1)$ it is readily found that $w^2 = k^2 c^2$, $c = w_p \lambda_D$. This of course is the solution of the dispersion relation (A3.5) in the limit $k \to 0$.

Importantly we now restrict attention to waves propagating in one direction only in which case we may write

$$\phi_1(x_1 t_1) = \sum_k \phi_k e^{ik(x_1 - ct_1)} .$$

It is then readily shown that $v_1 = e\phi_1/mc$ and $\rho_1 = -Ne\phi_1/k_B T$ and $\partial\phi_1/\partial t = -c\partial\phi_1/\partial x_1$.

By considering terms of $O(\varepsilon^5)$ one finds

$$\frac{\partial \rho_2}{\partial t_1} + N\frac{\partial v_2}{\partial x_1} = -\frac{\partial \rho_1}{\partial t_2} - \frac{\partial}{\partial x_1}(\rho_1 v_1)$$

$$\frac{\partial v_2}{\partial t_1} - \frac{e}{m}\frac{\partial \phi_2}{\partial x_1} = -v_1\frac{\partial v_1}{\partial x_1} - \frac{\partial v_1}{\partial t_2}$$

and

$$\rho_2 + Ne\phi_2/k_B T = (N/2)(e\phi_1/k_B T)^2 + \frac{\partial^2 \phi_1}{\partial x^2}/4\pi e .$$

Elimination of v_2 and ρ_2 gives

$$\left(\frac{\partial^2}{\partial x_1^2} - \frac{1}{c^2}\frac{\partial}{\partial t_1^2}\right)\phi_2 = \frac{1}{c^2}\frac{\partial}{\partial t_1}\left\{2\frac{\partial\phi_1}{\partial t_2} - \frac{e\phi_1}{mc}\frac{\partial\phi_1}{\partial x_1} + \frac{c^3}{w_p^2}\frac{\partial^3\phi_1}{\partial x_1^3}\right\} .$$

All the terms on the right-hand side are functions of $x_1 - ct_1$ and thus would produce secular behaviour (that is ϕ_2 having a contribution proportional to $x_1 + ct_1$). To avoid this we impose the condition

$$2\frac{\partial\phi_1}{\partial t_2} - \frac{e\phi_1}{mc}\frac{\partial\phi_1}{\partial x_1} + \frac{c^3}{w_p^2}\frac{\partial^3\phi_1}{\partial x_1^3} = 0 .$$

Similar equations hold for ρ_1 and v_1. It will be noted that all terms in this equation are of the same order in ε, namely ε^5. This justifies the original ordering.

Finally if we go to a coordinate frame moving with the phase velocity c, and put $\varepsilon = 1$, then we can replace x_1 by x and t_2 by t. The above is then the KdV equation.

Appendix A4:
The non-linear Schrödinger equation

An ideal case to consider would be the propagation of ion-acoustic waves for an arbitrary wave number. Unfortunately the necessary algebra becomes a little complicated and this distracts from the simplicity of the method. Thus, somewhat perversely perhaps, we consider the KdV equation as the model equation. Though this equation strictly applies only in the long-wavelength limit we assume here it applies for all wavelengths.

We use the multiple-scale perturbation theory. To lowest order, that is, linear theory, we consider a plane wave propagating with arbitrary wave number. The non-linear effects are weak, characterized by an expansion parameter ε, and they introduce a slow time and space modulation into the linear amplitude. These latter variations are characterized by the two times $\tau_1 = \varepsilon t$, $\tau_2 = \varepsilon^2 t$ and the space scale $x_1 = \varepsilon x$. Thus we assume an expansion of the solution of the KdV equation (6.8) of the form

$$u = u_0 + \varepsilon \ \{u_1(x_1, \tau_1, \tau_2) e^{i\theta} + \text{c.c}\} + O(\varepsilon^2) \ , \tag{A4.1}$$

where c.c. denotes complex conjugate and $\theta = kx - w(k)t$ with w a solution of the linear dispersion relation, namely

$$w = -6u_0 k - k^3 \ . \tag{A4.2}$$

To next order, the multiple-time perturbation theory gives

$$\frac{\partial u_2}{\partial t} - 6 \ u_0 \ \frac{\partial u_2}{\partial x} + \frac{\partial^3 u_2}{\partial x^3} = 6 \ ik \ [u_1^2 e^{2i\theta} - \text{c.c.}]$$

$$- \left[\left(\frac{\partial u_1}{\partial \tau_1} - (6u_0 + 3k^2) \frac{\partial u_1}{\partial x_1} \right) e^{i\theta} + \text{c.c.} \right] \ .$$

The terms proportional to $e^{\pm i\theta}$ are resonant and hence their coefficients have to be equated to zero. This is simply done by demanding that u_1 is a function of ξ, where $\xi = \varepsilon(x + (6u_0 + 3k^2)t)$. Then $u_2 = -[(u_1^2/k^2)e^{2i\theta} + \text{c.c.}]$. To next order in the expansion, terms of order $\exp(3i\theta)$ and $\exp(i\theta)$ are obtained. The latter of course resonate and their coefficient must be put to zero. This leads to the consistency condition

$$\frac{\partial u_1}{\partial \tau_2} - 3ik\frac{\partial^2 u_1}{\partial \xi^2} - \frac{18i}{k}|u_1|^2 u_1 = 0 \ .$$

Note from the definition of w, equation (A4.2), that $\xi \equiv \varepsilon(x - t\,dw/dk)$ and that the coefficient of $\partial^2 u_1/\partial \xi^2$ in the above equation is simply $(i/2)d^2w/dk^2$. Thus we write this equation in the form

$$i\frac{\partial u_1}{\partial \tau_2} + \frac{1}{2}\frac{d^2w}{dk^2}\frac{\partial^2 u}{\partial \xi^2} + \alpha|u|^2 u = 0 \ , \tag{A4.3}$$

where $\alpha = 18/k$. This is the non-linear Schrödinger equation.

This same sort of analysis may be applied to more general equations than the KdV. However, the final result is always an equation of the form (A4.3) with w defined appropriately for the particular problem and α also suitably defined. In particular for ion-acoustic waves, in the notation of Appendix A3,

$$w = w_p k\lambda_D/\sqrt{1 + k^2\lambda_D^2}$$

and

$$\alpha = -\left(\frac{w}{w_p}\right)^3 \frac{[3l^{10} + 6l^8 - 6l^6 - 29l^4 - 30l^2 - 12]}{12l^6(l^4 + 3l^2 + 3)} \ ,$$

where $l \equiv k\lambda_D$.

Appendix 5:
Fractal dimension

It is obvious from a perusal of a map of the world that some countries, Norway for example, have a rougher coastline than others, South Africa for example. What is not so obvious is how to quantify this roughness. Richardson in the 1920s made some suggestions and, at the turn of the century, Hausdorff considered what is essentially a similar problem though in a pure mathematics context. All this lay dormant until Mandelbrot brought such ideas together, introduced the concept of a fractal and showed the importance of fractals in science. The coastline of Norway is a fractal. However, no justice can be done in this book to the immense subject of fractals. Fortunately Mandelbrot (1982) has written a classic book and reference should be made to this for more details. Here it is sufficient to consider a few basic features, to introduce the concept of non-integer dimension and simple ways in which such a quantity may be calculated.

Richardson asked the question 'How long is the coastline of Britain?' and answered it by pointing out that it depends on the length of the ruler used in the measurement. Once pointed out this is obvious; by using a long ruler one misses out the small scale nooks and crannies. No matter how small the ruler there will be features on even a smaller scale. Of course this whole procedure must stop somewhere, as the concept of a coastline breaks down eventually, but over a range of sizes of rulers one can imagine the procedure. What Richardson also showed, by direct measurement of maps, was that if the length of the coastline measured with a ruler of length l was L, then

$$L \propto l^{\nu} \text{ or } \nu = \ln L/\ln l \ , \tag{A5.1}$$

where ν was a constant. The value of ν depends on the particular coastline and for Norway is of order 1.4 whilst for South Africa is of order 1.2. This led Richardson to suggest ν as a measure of the roughness of coastlines.

Suppose we want to calculate the length of a smooth curve such as a circle, using the above procedure. Since we consider using a straight ruler of length l say, the procedure is equivalent to measuring the perimeter of an inscribed polygon of side of length l. Consider a regular polygon of n sides each of length l. The perimeter of the polygon is equivalent to L so that we have $L = nl$ where $l = 2R\sin(\pi/n)$. Here R is the radius of the circle. Eliminating n gives $L = l\pi/\sin^{-1}(l/2R)$ which for small enough l gives the correct result $2\pi R$. This we can write in the form of A5.1 with $v = 0$. Since the geometric (topological) dimension of a circular line is 1 we write $v = d - 1$ where d is the dimension of the line. It is easy to imagine that if the line were so contorted that on the scale of the ruler it seemed to fill an area rather than a line, then with d now 2 we expect $v = 1$. Thus a non-integer value of v between zero and unity would seem to be a useful measure of the non-smoothness of a line. This is basically the idea behind fractal dimension.

To apply these ideas to complexity in phase space such as that found in strange attractors, we follow more closely the more mathematical concepts of Hausdorff. Consider an m-dimensional phase space. Of course m is integer. In the phase space the solution is represented as a series of lines or points. Now divide the phase space into m-dimensional boxes of side ε and calculate or use some numerical procedure to count the number of these elemental boxes which contain phase space trajectories or points. This number will of course depend on ε and we denote it by $N(\varepsilon)$. The Hausdorff dimension d_H is defined as

$$d_H = \lim_{\varepsilon \to 0} \frac{\ln N(\varepsilon)}{\ln \varepsilon} \; .$$

If the space were uniformly filled with trajectories then $N(\varepsilon) \propto \varepsilon^m$, which gives $d_H = m$, as one would expect. If the trajectories all lay on a smooth curve of dimension p (p an integer) then, since one only gets a contribution to N if the curve passes through a box, we get $N \propto \varepsilon p$, giving $d_H = p$. An example of a non-integer dimension is furnished by a strange attractor and is discussed more fully in Chapter 8.

Now, just as for the case of the coastline, which is always a line and hence one-dimensional, the value of d_H as defined above will be related to the roughness of the coast. Since we take ε to be l then $N(\varepsilon)$ must be of order 1 and we have from above that $d_H \simeq \ln L/\ln l = (v + 1)$.

In practice it is not possible in any calculation to go to the limit $\varepsilon = 0$, and what is usually found is that $\ln N(\varepsilon)$ plotted as a function of $\ln \varepsilon$ is a straight line over a range of values of ε. The slope of this line is then taken to be d_H. The range of ε can be over three or four orders of magnitude but it is usually found that significant deviations occur for the smaller values. Thus one uses the concept of dimension not so much in the strictly mathematical limit of $\varepsilon \to 0$ but for a range of values of ε with ε always finite.

Other definitions of dimension have been considered, because although the concept of Hausdorff dimension is relatively simple to define it is not particularly convenient to calculate in practice. The number of points in the phase space needed before one can obtain reasonable estimates of d_H increases with the square of the dimensions of the phase space. To alleviate such difficulties other definitions have been proposed and a particularly useful one is the correlation dimension.

Imagine iterating a one-dimensional map, such as the logistic equation (7.1), a large number of times, N say, and storing the values in increasing order from zero to one. Label these values as x_m with m ranging from zero to N. Note that x_m is not the mth iterate of the map. The correlation function $C(\Delta)$ is defined by

$$C(\Delta) = \lim_{N \to \infty} \frac{1}{N} \sum_{i \neq j}^{N} H \left(\Delta - |x_i - x_j| \right) ,$$

where H is the Heaviside function defined such that $H(x) = 1$ for $x > 0$ and $H(x) = 0$ for $x < 0$. The correlation dimension d_c is defined such that for $\Delta \to 0$

$$C(\Delta) \simeq \Delta^{d_c} .$$

It has been shown that $d_c \leqslant d_H$.

The meaning of $C(\Delta)$ is made clear by calculating it using the above definition. A value for Δ is chosen together with a value of x, x_i say. One then computes the number of values within a distance of Δ of this point, divides by the total number of values, N, so as to obtain an equivalent density. This density, which depends on x_i, is then averaged over all values to give $C(\Delta)$. Thus $C(\Delta)$ is just the average density of points in an element of length Δ. With this interpretation the analogous definition of $C(\Delta)$ in an M-dimensional phase space is simply the average density of points of an M-dimensional sphere of radius Δ.

When applying these ideas to an experimental situation one does not in general know M and usually one only knows the time evaluation of one quantity $y(t)$ say. To apply the above ideas one starts by assuming a value for M and then constructing the vectors $\mathbf{x}_n$ with components x_n^j at discrete times $t = n\tau$ such that

$$x_n^j = y \left(ns + j\tau \right) , \qquad 0 \leqslant j \leqslant M ,$$

where s is a chosen integer. $C(\Delta)$ is calculated, and hence d_c, for increasing values of M, and if d_c converges, this value is identified with the dimension of the attractor. Because of limits on the available amount of experimental data the above method is useful only if $d_c \leqslant 6$.

Questions on the appendices

1. Prove that the adjoint operation corresponding to the operator L defined in (A1.2) is as given by (A1.3).
2. Apply the multiple-scale perturbation theory to the equation for the anharmonic oscillator

$$\frac{d^2x}{dt^2} + w^2x = \varepsilon x^3 \ ,$$

   where $\varepsilon \ll 1$. Obtain a periodic solution for x correct to $O(\varepsilon^2)$.
3. Apply multiple-scale perturbation theory to the equation

$$\frac{d^2x}{dt^2} + w^2x = \varepsilon x^2$$

   and obtain the periodic solution to $O(\varepsilon^3)$.
4. Use multiple-scale perturbation theory to reduce the equations

$$\frac{\partial h}{\partial t} + \frac{\partial}{\partial x}(hu) = 0$$

   and

$$\frac{\partial u}{\partial t} + u\frac{\partial u}{\partial x} = -g\frac{\partial h}{\partial x} - \alpha\frac{\partial^3 h}{\partial x\partial t^2}$$

   to a KdV equation.

Answers

CHAPTER 2

1. The condition for a turning point is $dF/dM = 0$ which from (2.4) gives either $M = 0$ or $2\alpha(T - T_c) + 3\gamma M + 4\beta M^2 = 0$. From the latter equation we have

$$M_{\pm} = (-3\gamma \pm \sqrt{9\gamma^2 - 32\alpha\beta(T - T_c)})/8\beta \ ,$$

and this has real solutions (since M is a measurable physical quantity it must be real) if

$$9\gamma^2 > 32\alpha\beta(T - T_c) \ .$$

The condition for a maximum (minimum) is that $d^2F/dM^2 < 0(>0)$ and this gives that the turning point $M = 0$ is a minimum for $T > T_c$ and maximum for $T < T_c$.

The expression for d^2F/dM^2 can be simplified by using the form of the equation and the solution for the other two fixed points. Thus

$$\frac{d^2F}{dM^2} = 2\alpha(T - T_c) + 6\gamma M + 12\beta M^2 \ ,$$

which reduces to

$$= -4\alpha\Delta T - 3\gamma M \ ,$$

if the equation for M is used to remove the M^2 term. Then using the above form for M we find

$$\frac{\mathrm{d}^2 F}{\mathrm{d}M^2} = [\sqrt{9\gamma^2 - 32\alpha\beta(T - T_\mathrm{c})} \pm 3\gamma] \sqrt{9\gamma^2 - 32\alpha B(T - T_\mathrm{c})} \ .$$

The turning point associated with the $+$ sign (M_-) is thus always a minimum whilst that associated with the negative sign (M_+) changes from a maximum for $T > T_\mathrm{c}$ to a minimum for $T < T_\mathrm{c}$.

2. For $T > T_\mathrm{c}$ the minima occur for $M = 0$ and M_-. Then the condition that $F(0) = F(M_-)$ gives $\alpha(T - T_\mathrm{c}) + \gamma M_- + \beta^2 M_- = 0$. Using the equation for M this condition simplifies to $\gamma M_- = -2\alpha(T - T_\mathrm{c})$ and finally using the expression for M_- the condition $T - T_\mathrm{c} = \gamma^2/4\alpha\beta$ is obtained. For $T < T_\mathrm{c}$ the minimum points are M_+ and M_- so the condition becomes $F(M_+) = F(M_-)$. Repeated use of the equation for M shows that we may write

$$F(M) = \bar{F} - [9\gamma^2 - 32\alpha\beta(T - T_\mathrm{c})]\gamma M/256\beta \ ,$$

where $\bar{F}$ does not depend on M. Thus no solution can exist where $F(M_+) = F(M_-)$ since $9\gamma^2 - 32\alpha(T - T_\mathrm{c}) > 0$. This whole behaviour is consistent with the sketches of $F(M)$ shown in Fig. (2.3).

3. The major difference between the expression $\bar{V}(p)$ and $V(q)$ as given by (2.5) is that no term proportional to q^3 occurs in (2.5). Thus the first step is to remove the p^3 term in $\bar{V}(p)$. This is done by writing $p = p_1 + \alpha$, substituting into the expression for $V(p)$ and demanding that the coefficient of p_1^3 is identically zero. This gives $\alpha = -a_3/4a_4$ and

$$\bar{V} = a_4 p_1^4 + (a_2 - (3a_3^2/8a_4))p_1^2 + (a_1 - (2a_2 a_3/4a_4) + (a_3^2/8a_4^2))p_1 + \text{constant} \ .$$

(The constant term is of no relevance as it is the derivatives of $\bar{V}$ which control the behaviour of the system.) Finally by identifying p_1 with $q/(4a_4)^{1/4}$ the expression $\bar{V}(p)$ may be made identical to equation (2.5) with the identification

$$u = (a_2 - (3a_3^2/8a_4))/(4a_4)^{1/2}$$

and

$$v = (a_1 - (2a_2 a_3/4a_4) + (a_3^2/8a_4^2))/(4a_4)^{1/4} \ .$$

Thus the form of potential given by (2.5) is the most general form of potential which includes all powers of the order parameter up to and including a quartic term. However, it should be borne in mind that in practice it is the as which are directly related to the physical world and not the μ and v.

4. Using the identity that for $x < 1$

$$\ln(1-x) = -\left(x + \frac{x^2}{2} + \frac{x^3}{3} + \frac{x^4}{4} + \ldots\right) ,$$

the Van der Vaal free energy may be expanded in powers of M to give

$$F = M^2(bT-a)/2 + b^2TM^3/6 + b^3TM^4/12 .$$

Comparison with (2.4) allows one to identify T_c with a/b. Note that the terms of the other powers of M are all positive and finite in the limit $T \to T_c$, a requirement demanded in the analysis of equation (2.4) given in Chapter 2.

5. Differentiation of the Van der Vaal free energy gives

$$\frac{dF}{dM} = -T\ln(1-bM) - (a/b)bM ,$$

so the extremal satisfies the equation

$$T\ln(1-bM) = -(a/b)bM .$$

This equation for M cannot be solved in closed form. However sketching both sides of this equation as a function of bM shows that for $T > (a/b)$ there is only one intersection and this is at $M = 0$. (Note that for small M the term $\ln(1-bM)$ is of order $-bM$.) For $T < (a/b)$ there are three intersections, including one at $M = 0$.

The point of inflexion is defined such that $d^2F/dM^2 = 0$ which in this case gives $1 - bM = Tb/a$. Substituting this expression for M into the condition $dF/dM = 0$ gives the following equation for the temperature

$$t - t\ln t = 1 ,$$

where $t = Tb/a$. By inspection it is seen that a solution to this equation is $t = 1$. Again by sketching the left-hand side as a function of t this is found to be the only solution.

If the form for F obtained in the answer to question 4 is now taken the conditions $dF/dM = 0$ and $d^2F/dM^2 = 0$ give

$$1 - \frac{1}{t} + \frac{bM}{2} + \frac{(bM)^2}{3} = 0$$

and

$$(1 - 1/t) + bM + (bM)^2 = 0 \ .$$

Subtracting these equations gives $bM = -3/4$, which, when substituted into either equation, gives the condition $t = 16/13$.

CHAPTER 3

1. By direct differentiation and use of equation (3.1) for q one has

$$\frac{d\phi}{dt} = 2q \frac{dq}{dt} = 2q \ (\alpha q - \beta q^3) = 2\alpha\phi - 2\beta\phi^2 \ ,$$

an equation for ϕ.

The solution of this first-order equation is given by

$$\int \frac{d\phi}{(\alpha - \beta\phi)\phi} = 2(t + t_0) \ ,$$

where t_0 is a constant of integration. By expressing the integrand in terms of partial fractions the integration may be carried out to give

$$\frac{1}{\alpha} \ln|\phi| - \frac{1}{\alpha} \ln(|\alpha - \beta\phi|) = 2(t + t_0) \ ,$$

so that

$$\ln \ (|\phi|/|\alpha - \beta\phi|) = 2\alpha(t + t_0) \ .$$

This may be written in the form

$$|\phi|(|\alpha - \beta\phi|) = |\phi_0|e^{2\alpha t}/|\alpha - \beta\phi_0| \ ,$$

where ϕ_0 is the value of ϕ at $t = 0$ ($|\phi_0|/|\alpha - \beta\phi_0| \equiv e^{2\alpha t_0}$). Simple algebra now gives

$$\phi = \alpha\phi_0 e^{2\alpha t}/(\alpha + \beta\phi_0(e^{2\alpha t} - 1)) \ .$$

This is true for either sign of $\alpha - \beta\phi_0$. If $\phi_0 > 0$ then for $\alpha t \to \infty$, $\phi \to \alpha/\beta$ independent of the initial condition ϕ_0 whilst for $\alpha t \to -\infty$, $\phi \to 0$ if $\alpha - \beta\phi_0 > 0$.

However, if $\phi_0 < 0$ a time will be reached when the denominator becomes zero and ϕ becomes infinite. Such behaviour where the solution of an equation, in this case ϕ, becomes infinite in a finite time is called an explosive instability. In the present case this finite time is equal to $\ln((\alpha - \beta\phi_0)/\beta\phi_0)$.

2. The time-independent states $\overline{q}$ satisfy the equation $\overline{q}(\alpha - \beta\overline{q}) = 0$, so that there are always two distinct values $\overline{q} = 0$ and $\overline{q} = \alpha/\beta$. Writing $q(t) = \overline{q} + \delta q(t)$ and substituting into the differential equations gives

$$\frac{d\delta q}{dt} = (\alpha - 2\beta\overline{q})\delta q - \beta\,\delta_q^2\ .$$

This is exact and equivalent to the original differential equation. We now linearize by neglecting the term proportional to δq^2 in which case the equation may be solved to give

$$\delta q = \delta q_0 \exp((\alpha - 2\beta\overline{q})t)\ ,$$

where δq_0 is the value of $\delta q(t)$ at $t = 0$. Thus the stability condition, namely that $\delta q \to 0$ as $t \to \infty$, demands that $\alpha - 2\beta\overline{q} < 0$. Thus the state $\overline{q} = 0$ is stable if $\alpha < 0$ whilst the state $\overline{q} = \alpha/\beta$ is stable if $\alpha > 0$.

Now the equation is of the form of (3.7) but with $H(q) = \alpha q - \beta q^2$. Since $H(q) > 0$ for $0 < q < \alpha/\beta$ and $H(q) < 0$ for all $q > \alpha/\beta$ only initial values of q which are greater than zero will be attracted to the fixed point at $\overline{q} = \alpha/\beta$. However, $H(q) < 0$ for all $q < 0$ and hence any initial value which is less than zero will increase indefinitely.

The differential equation for q is the same as that for ϕ discussed in question 1. Comparison of the two forms of answers obtained for essentially the same equation shows that the explosive instability can be identified with the fact that $H(q)$ monotonically decreases for $q < 0$.

3. The time-independent solutions to the equation occur for $\sin\overline{q} = 0$, that is $\overline{q} = n\pi$ where n is any integer. Writing $q(t) = n\pi + \delta q(t)$, substituting into the equation and neglecting terms of $O(\delta q^2)$ gives

$$\frac{d\delta q}{dt} = (\cos n\pi)\,\delta q\ .$$

Thus the fixed points $\overline{q} = (2m + 1)\pi$ are stable whilst those where $\overline{q} = 2m\pi$ are unstable (m any integer).

The global features of the solution may be studied by identifying $H(q)$ in (3.7) with $\sin q$. Since for $2m\pi < q < (2m + 1)\pi$, $H(q) > 0$ all initial values of q in this range are attracted to the end point $q = (2m + 1)\pi$. Similarly it is found that since for $(2m + 1)\pi < q < 2(m + 1)\pi$, $H(q) < 0$ all initial points in this range are attracted to the end point $(2m + 1)\pi$. In both cases the solution eventually ends up at the same fixed point which linear stability analysis showed to be stable.

4. Directly from the differential equation we may write

$$t + t_0 = \int^q \frac{dq}{\sin q} = \ln|\tan(q/2)| \ ,$$

so that

$$q = 2 \tan^{-1} (\exp(t + t_0)) \ .$$

The constant t_0 is now chosen such that $q = q_0$ at $t = 0$. This gives $q_0 = 2 \tan^{-1}(\exp(t_0))$ so that

$$q = 2 \tan^{-1}(\tan(q_0/2)e^t)$$

For $t \to \infty$ the argument of the $\tan^{-1}$ function goes to positive or negative infinity depending on the sign of $\tan(q_0/2)$. Thus since for $2m\pi < q_0 < (2m + 1)\pi$, $\tan(q_0/2) > 0$ then for $t \to \infty$, the argument of the $\tan^{-1}$ is positive and $q \to (2m + 1)\pi$. Similarly for $(2m + 1)\pi < q_0 < 2(m + 1)\pi$ the argument is negative and $q \to (2m + 1)\pi$ again. This is of course in complete agreement with the results obtained for question 3 by using global analysis.

5. The stochastic equation appropriate to this deterministic equation is (3.16) with $H(q) = -a\tanh q$. Then the Fokker–Planck equation (3.18) takes the particular form

$$\frac{\partial P}{\partial t} - a \frac{\partial}{\partial q} (\tanh q \ P) = D \frac{\partial^2 P}{\partial q^2} \ .$$

The time-independent form is obtained by neglecting the $\partial P/\partial t$ term. The resulting equation is readily integrated to give

$$D \frac{dP}{dq} = -a\tanh q \ P(q) \ .$$

A constant of integration has been put identically zero to ensure that $P(q)$ goes to zero as $q \to \pm \infty$. For constant D this equation may be integrated once more to give

$$P(q) = P_0\, e^{\frac{-a}{D}\ln(\cosh q)} = \frac{P_0}{(\cosh q)^{a/D}}\ ,$$

where P_0 is a constant. This constant is obtained by demanding that $P(q)$ is normalized to unity so that

$$1 = P_0 \int_{-\infty}^{+\infty} \frac{dq}{(\cosh q)^{a/D}}\ .$$

The integral can be expressed for any value of a/D in terms of beta functions but for the particular case of $a/D = 2$ it simplifies to give $P_0 = \frac{1}{2}$. Then the unique expression for $P(q)$ is

$$P(q) = \mathrm{sech}^2(q)/2\ .$$

In the general case it is readily seen that $dP/dq = 0$ for $q = 0$ and further d^2P/dq^2 at $q = 0$ is less than zero so that $P(q)$ is maximum at $q = 0$.

The global features of the solution of the deterministic equation are readily found. Since $\tanh q > 0$ for all $q > 0$ any initial condition q_0 which is greater than zero will evolve with time towards the origin. Similarly any initial condition less than zero evolves towards the origin since $\tanh q < 0$ if $q < 0$. The origin $q = 0$ is the only stable point. Thus the maximum of the probability function $P(q)$ coincides with the only stable point of the deterministic equation.

6. For this particular form for $H(q)$, equation (3.23), for the mean first-passage time M, takes the form

$$\frac{1}{2}\frac{d^2M}{dq^2} + \tanh q\,\frac{dM}{dq} = -1\ .$$

This is really a first-order equation for dM/dq and may be solved for this quantity by introducing the integrating factor $S(q)$ where $dS/dq = 2S\tanh q$ that is $S(q) = \cosh^2 q$. Then the differential equation may be written in the form

$$\frac{d}{dq}\left(S\frac{dM}{dq}\right) = -2S\ .$$

This may be integrated once to give

$$\frac{dM}{dq} = -\left(q + \frac{1}{2}\sinh2q\right)/\cosh^2q$$

and integrated again to give

$$M(q) = M_0 - q\,\tanh q\ ,$$

where M_0 is a constant of integration. Substitution of this solution into the original differential equation confirms the accuracy of the analysis.

We now impose the boundary condition that $M = 0$ for $q = \pm1$. This gives that $M_0 = \tanh1$ so that the final unique value of M is given by

$$M(q) = \tanh1 - q\tanh q\ .$$

In obtaining this result a constant of integration that arises after the first integration was put to zero. Had it been included the application of the boundary conditions would have shown it to be zero. A little problem for the student.

CHAPTER 4

1. Equation (4.1) can be expressed in the standard form, as given by equations (4.4) and (4.5), by taking $\theta = x$ and $d\theta/dt = y$. Then $dy/dt(= d^2\theta/dt^2) = -(g/l)\sin x$ and so $H(x,y) = y$ and $R(x,y) = -(g/l)\sin x$. The fixed points are $\bar{y} = 0$ and $\sin\bar{x} = 0$, that is $\bar{x} = n\pi$, with n any integer. Now writing $y = \delta y(t)$, $x = \bar{x} + \delta x(t)$, substituting into the equations of motion and neglecting terms of $O(\delta x^2)$ gives (note that to this approximation $\sin(x(t)) = \sin(\bar{x}) + \cos(\bar{x})\delta x(t) = (-1)^n\delta x$)

$$\frac{d\delta x}{dt} = \delta y$$

and

$$\frac{d\delta y}{dt} = -(-1)^n(g/l)\delta x\ .$$

Asssuming a solution of the form $\delta x = Ae^{\lambda t}$ gives from the first equation that $\delta y = \lambda Ae^{\lambda t}$, which when substituted into the second equation gives

$$(\lambda^2 + (g/l)(-1)^n)Ae^{\lambda t} = 0\ .$$

For this equation to be satisfied for a range of values of t (excluding the trivial solution $A = 0$ which gives $\delta x = \delta y = 0$) we must have

$$\lambda^2 + (g/l)(-1)^n = 0 .$$

This result could have been obtained directly fron the indicial equation (4.9) since $\alpha = \eta = 0$, $\beta = 1$ and $\gamma = -(-1)^n(g/l)$.

For n even we have $\lambda = \pm i\sqrt{g/l}$ and the corresponding fixed points are centres.

For n odd we have $\lambda = \pm \sqrt{g/l}$ and the fixed points are saddles.

The complete solution of the linearized equations near a saddle point take the form

$$\delta x = Ae^{\eta t} + Be^{-\eta t}$$

and $(\delta y = d\delta x/dt)$

$$\delta y = \eta Ae^{\eta t} - \eta Be^{-\eta t} ,$$

where $\eta^2 = g/l$. For large values of t the terms proportional to B may be neglected, in which case $\delta y = \eta \delta x$.

Similarly for $t \to -\infty$, $\delta y = -\eta \delta x$. The lines defined by $\delta y = \pm \eta \delta x$ define the separatrix.

In the vicinity of the centre it is convenient to write $\delta x = A\cos(\eta t + \phi)$ with A and ϕ the two arbitrary constants. Then $\delta y = -\eta A\sin(\eta t + \phi)$ so that

$$\delta x^2 + \delta t^2/\eta^2 = A^2$$

and the contours in the phase plane are simple ellipses, different ellipses corresponding to different values of A.

Note that the contours in the phase plane give no indication of where the solution is at a particular time, merely where the solution is confined to roam for all time. Also that the amplitude scaling in the two perpendicular directions δx, δy, namely η, is equal to $2\pi/T$, where T is the period of the motion.

2. The basic equation is now

$$\frac{d^2\theta}{dt^2} + a\frac{d\theta}{dt} + (g/l)\sin\theta = 0$$

and this can be put in the standard form of (4.4) and (4.5) with $\theta = x$, $d\theta/dt = y$, $H = y$ and $R = -(g/l)\sin x - ay$. Following the procedure given in the text and also in more detail in question 1, we find the fixed points are still at $\bar{y} = 0$ and $\bar{x} = n\pi$ but the indicial equation is now $(\eta^2 = g/l)$

$$\lambda^2 + a\lambda + \eta^2(-1)^n = 0 \ .$$

This equation has two roots λ_+, λ_- given by

$$2\lambda_\pm = -a \pm \sqrt{a^2 - 4\eta^2(-1)^n} \ .$$

For the case $a/\eta \ll 1$ the square root term may be expanded to give to $O(a^2)$

$$2\lambda_\pm = -a \pm 2\eta\sqrt{-(-1)^n} \ .$$

For n odd the inclusion of the damping term produces a quantitative change only. The fixed points remain saddles with separatrices now defined by $\delta y = (\eta - a/2)\delta x$ and $\delta y = -(\eta + a/2)\delta x$, that is a simple change of slope. In fact as can be seen from the definition of $\lambda_\pm$ these fixed points remain saddles for all values of a. ($\lambda_\pm$ are always real and of opposite sign.)

For n even the situation is very different. With $a/\eta \ll 1$ we see that

$$\lambda_\pm = \pm i\eta - a/2$$

and this corresponds to a stable focus. The centre ($a \equiv 0$) now undergoes a qualitative change and becomes a stable focus.

The time dependence of $\delta x(t)$ can now be written in the form

$$\delta x(t) = A e^{-at/2}\cos(\eta\, t + \phi) \ ,$$

corresponding to damped harmonic motion and to $O(\delta a)$

$$(\delta x)^2 + (\delta y/\eta)^2 = A^2 e^{-at} \ .$$

The orbit follows an elliptical path which spirals towards the fixed points at $\theta = 2n\pi$, $d\theta/dt = 0$.

The above gives an example of the concept of structural stability. The saddle remains a saddle even after a change in the form of the equation whilst the centre does not. Saddles are structurally stable, centres are not. Physically this means that the introduction of damping, no matter how small, will mean that an oscillating pendulum will eventually come to rest ($\theta = 0$, $d\theta/dt = 0$), but no matter how large the damping an inverted pendulum ($\theta = \pi$, $d\theta/dt$) will always fall.

For $a > 2\eta$ the n even fixed points become stable nodes ($\lambda_+ < 0$, $\lambda_- < 0$). In the pendulum case this is referred to as the overdamped situation.

3. Direct differentiation with respect to t of the expression given in the question gives

$$\frac{dR}{dt}(\eta/R - B) - \frac{dF}{dt}(a - \lambda/F) = 0 \ .$$

Using equation (1.7) to substitute for dR/dt and dF/dt shows that the above is automatically satisfied.

An alternative approach is to solve (1.7) directly. To this end we consider $R(t)$ to be a function of t through $F(t)$ only. That is $R(t) = R(F(t))$ so that $dR/dt \equiv (dR/dF)dF/dt$. Then using (1.7) we have

$$dR/dF = (dR/dt)/(dF/dt) = -\frac{(\lambda - \alpha F)R}{(\eta - \beta R)F} \ .$$

The assumption is now seen to be justified since dR/dF is a function of F only and t no longer enters the equation. This equation may be rearranged and integrated to give

$$\int^{R} \frac{dR}{R}(\eta - \beta R) = -\int^{F} \frac{dF}{F}(\lambda - \alpha F) + \text{constant} \ .$$

The integrals are all straightforward and the result given in the question is readily obtained.

In principle one could solve this relationship and express R as a function of F only, $R = G(F)$ say. Then using the second of the equations in (1.7) we have

$$\frac{dF}{dt} = -(\eta - \beta G(F))F \ ,$$

which itself may be solved to give

$$\int^{F} \frac{dF}{F(\eta - \beta G(F))} = -t + \text{constant} \ .$$

The problem is then said to have been reduced to quadrature, that is it involves the evaluation of a one-variable integral.

Unfortunately it is not possible to obtain an analytical form for $G(F)$ and so the above procedure of obtaining an explicit form for the time variation of F is also not possible.

Of course one can solve for $G(F)$ numerically and in fact this was the procedure used to produce Fig. 4.6. The two curves in this figure correspond to different values of C.

4. Letting $y = dx/dt$ this equation can be written in the standard form of equations (4.4) and (4.5) with $H = y$ and $R = -\omega^2 x - yf(x)$. The only fixed point is $x = 0$, $y = 0$. Linearizing about this point gives

$$\frac{d\delta x}{dt} = \delta y$$

and

$$\frac{d\delta y}{dt} = -\omega^2 \delta x - f(0)\delta y \ .$$

The indicial equation (4.9) reduces to

$$\lambda^2 + f(0)\,\lambda + \omega^2 = 0 \ ,$$

so that

$$2\lambda_\pm = -f(0) \pm \sqrt{f^2(0) - 4\omega^2} \ .$$

Now a necessary condition for a stable limit cycle is that orbits in the vicinity of the origin spiral outwards. That is, the fixed point at the origin must be an unstable focus. From the form given for $\lambda_\pm$ this condition demands that $f(0) < 0$ *and* $4\omega^2 > f^2(0)$. That is the λs are complex with a positive real part.

It is important to note that this is only a necessary condition. For the case of the Van der Pol equation (4.11), $f(x) = -a(1 - x^2)$ so that $f(0) = -a$. The origin is thus an unstable focus for $2\omega_0 > a > 0$. In this case other considerations show that a limit cycle exists. In particular for large x ($x^2 > 1$) the coeffficient of dx/dt changes sign.

Unfortunately from an application point of view obtaining both necessary and sufficient conditions for the existence of a limit cycle is a difficult problem.

5. From the definition of a potential namely $F(x) = -\partial V/\partial x$, simple integration shows for the particular case of this question that

$$V(x) = ax \text{ for } x > 0$$

$$= -ax \text{ for } x < 0 \ ,$$

or $V(x) = a|x|$ all x. Then from (4.14) we have, for $x > 0$, that

$$\int \frac{dx}{\sqrt{K - ax}} = \sqrt{2}(t + t_0) \ ,$$

where t_0 is the constant of an integration. The integral is readily evaluated to give

$$-\frac{2}{a}\sqrt{K - ax} = \sqrt{2} \ (t + t_0) \ ,$$

or

$$x = K/a - a(t + t_0)^2/2 \ .$$

A similar analysis for $x < 0$ gives

$$x = -K/a + a(t + t_1)^2/2 \ ,$$

where t_1 is a different integration constant. To construct the complete solution one proceeds in the following manner. Suppose at $t = 0$, $x(0) = x_0 > 0$. Then $t_0 = \sqrt{2(K/a - x_0)}/a$ and thus we must impose the condition that $x_0 < K/a$. Now as t increases x increases until it reaches its maximum value K/a ($t = -t_0$), changes direction and eventually reaches the origin at $t = -t_0 + \sqrt{K/a^2}$. For larger values of time one must use the expression for $x(t)$ for $x(t) \leqslant 0$ with $t_1 = -t + \sqrt{2K/a^2}$. With further increase in time $x(t)$ decreases until it reaches its minimum value at $-K/a$. Afterwards $x(t)$ undergoes periodic motion, oscillating between $\pm K/a$.

To calculate the period of this motion imagine that at $t = 0$, $x = 0$ so $t_0 = 0$ then the time to go from $x = 0$ to $x = K/a$ and return to $x = 0$ is equal to

$$2\int_0^{K/a} \frac{dx}{\sqrt{K - ax}} = 2\sqrt{K}/a \ .$$

Thus the period of oscillation is $4\sqrt{K}/a$, the extra factor 2 arising because the solution goes negative before returning to the origin again.

CHAPTER 5

1. Comparison of equations (3.1) and (5.5) shows that $\alpha \equiv \lambda_1$ and $\beta \equiv ab/\lambda_2$ so that the complete solution of equation (5.5) takes the form

$$q_1^2 = q_0^2\, e^{2\lambda_1 t}/[1 - (abq_0^2/\lambda_1\lambda_2)(1 - e^{2\lambda_1 t})]\ ,$$

where q_0 is the value of $q_1(t)$ at $t = 0$. This may be written in the form

$$q_1^2 = \frac{(\lambda_1\lambda_2/ab)}{1 + (\lambda_1\lambda_2/abq_0^2 - 1)e^{-2\lambda_1 t}}\ ,$$

which for large time $(e^{-2\lambda_1 t} \ll 1)$ may be expanded to give

$$q_1^2 = (\lambda_1\lambda_2/ab)[1 - (\lambda_1\lambda_2/abq_0^2 - 1)e^{-2\lambda_1 t}] + O\ (e^{-4\lambda_1 t})\ .$$

Using the identity that $\sqrt{1 + \varepsilon} \simeq 1 + \varepsilon/2$ for $\varepsilon \ll 1$ finally gives

$$q_1(t) = \sqrt{\lambda_1\lambda_2/ab}\ \{1 - \tfrac{1}{2}(\lambda_1\lambda_2/abq_0^2 - 1)e^{-2\lambda_1 t}\}\ .$$

Thus we may make the following identification

$$q = \sqrt{\lambda_1\lambda_2/ab}\ \text{ and }\ B = \sqrt{\lambda_1\lambda_2/4ab}\ (1 - \lambda_1\lambda_2/abq_0^2)\ .$$

Inspection of the exact solution obtained above shows that if $\lambda_1\lambda_2/abq_0^2 = 1$ then $q_1(t) = \lambda_1\lambda_2/ab$ for all time. That is $q_1(t)$ is time-dependent. Note that this particular choice of q_0 corresponds to the fixed point of (5.5).

2. Since $\lambda \gg 1$ we expect that the value of q_2 will relax very quickly to its equilibrium value obtained by putting $dq_2/dt \equiv 0$. This gives $q_2 = q_1$ so that q_2 is directly enslaved by q_1. Substitution of this expression into the first equation gives

$$\frac{dq_1}{dt} = q_1 + \sqrt{a^2 + q_1^2}\ .$$

This equation may be solved by writing

$$\int \frac{dq}{q + \sqrt{a^2 + q^2}} = t + t_0\ ,$$

where t_0 is the constant of integration. To evaluate the integral we make the substitution $q = a\sinh\theta$ to give

$$\frac{1}{2} \int (1 + e^{-2\theta}) \, d\theta = t + t_0 \; .$$

The integrals are now straightforward and one finally obtains the implicit solution $2\theta - e^{-2\theta} = 4(t + t_0)$ which is equivalent to

$$2 \sinh^{-1} (q_1/a) - (\sqrt{1 + q_1^2/a^2} - q_1/a)^{1/2} = 4(t + t_0) \; .$$

For $t \rightarrow +\infty$ one must also have q_1/a getting large, then for $q_1/a \gg 1$ the left-hand side of the above equation simplifies to $(\sinh^{-1}x \simeq \ln 2x)$ $2\ln(2q_1/a)$ so that $2\ln(2q_1/a) \simeq 4t$ or

$$q_1 \simeq (a/2)e^{2t} \; .$$

3. By introducing a scaling and writing $C = P_o\bar{C}$, $x = P_1\bar{x}$ and $t = P_2\bar{t}$, where P_0, P_1 and P_2 are constants, substituting into (5.6) and dividing by P_0/P_2 this equation reduces to

$$\frac{\partial \bar{C}}{\partial t} = \alpha P_2\bar{C} - \beta \, P_0^2 P_2\bar{C}^3 + (DP_2/P_1^2) \frac{\partial^2 \bar{C}}{\partial \bar{x}^2} \; .$$

If we now make the identification $\alpha P_2 = 1$, $\beta P_0^2 P_2 = 1$ and $DP_2/P_1^2 = 1$ the equation reduces to a form which does not involve any constants, namely

$$\frac{\partial \bar{C}}{\partial \bar{t}} = \bar{C} - \bar{C}^3 + \frac{\partial^2 \bar{C}}{\partial \bar{x}^2} \; .$$

The three algebraic equations are readily solved for P_0, P_1 and P_2 so finally we may write

$$C(x,t) = \sqrt{\alpha/\beta} \; \bar{C}(\sqrt{\alpha/D} \, x, \alpha t) \; .$$

The advantages of this scaling procedure are now obvious. All solutions to equations of the form (5.6) are simply related to the universal equation given above for $C(x,t)$.

However, when one has to impose boundary conditions it is necessary to introduce one dimensionless parameter. The boundary condition $\partial C/\partial x = 0$ at $x = 0$ and $x = L$ is easily seen to imply $\partial \bar{C}/\partial \bar{x} = 0$ at $\bar{x} = 0$ and $\bar{x} = L/P_1$, that is at $\bar{x} = 0$ and $\bar{x} = \bar{L}$, where the one parameter remaining in the problem is

$\overline{L} = L\sqrt{D/\alpha}$. Note that the parameter β does not enter and only serves to control the magnitude of $C(x,t)$.

The scaled version of (5.7) may be obtained directly from the above equation for $\overline{C}$ or by introducing the scaling into (5.7). Either way results in an equation of the form (5.7) but now in terms of $\overline{A}$ and $\overline{\tau}$, with $\beta \equiv 1$ and $\lambda_n = (n\pi/\overline{L})^2 - 1$ and of course $A = \sqrt{\alpha/\beta} \, \overline{A}$.

4. The equations are linear and have constant coefficients. Hence one looks for solutions of the form $q_1 = A \exp(\eta t)$ and $q_2 = B \exp(\eta t)$ with A, B and η independent of time. Substitution of these forms into the equations gives, if $A \neq 0$, $B \neq 0$, the condition

$$\eta^2 + (\lambda_2 - \lambda_1)\eta - (ab + \lambda_1\lambda_2) = 0$$

and $B = A(\eta - \lambda_1)/a$. Thus the complete solution to the set of equations is

$$q_1 = A_+ e^{\eta_+ t} + A_- e^{\eta_- t}$$
$$q_2 = (\eta_+ - \lambda_1)(A_+/a)e^{\eta_+ t} + (\eta_- - \lambda_1)(A_-/a)e^{\eta_+ t} \, ,$$

where

$$2\eta_\pm = -\lambda_2 \{1 - (\lambda_1/\lambda_2 \pm \sqrt{1 + 2\lambda_1/\lambda_2 + (\lambda_1^2 + 4ab)/\lambda_2)^2}\} \, .$$

Now for $\lambda_1/\lambda_2 \ll 1$ we may expand in powers of this ratio and obtain $\eta_+ \simeq -\lambda_2$ and $\eta_- \simeq \lambda_1 + ab/\lambda_2$. Note, to get this last result, one must use the identity $\sqrt{1 + \varepsilon} \simeq 1 + \varepsilon/2 - \varepsilon^2/8 + O(\varepsilon^3)$. In this case for sufficiently long times $\lambda_2 t \gg 1$ we may neglect the term proportional to $\exp(\eta_+ t)$ and then

$$q_1(t) \simeq A_- \exp((\lambda_1 + ab/\lambda_2)t)$$

and

$$q_2(t) \simeq b \, q_1(t)/\lambda_2 \, .$$

If we invoke the enslaving principle then, since $\lambda_2 \gg \lambda_1$, the second equation gives $q_2 = bq_1/\lambda_2$, a result in complete agreement with the exact solution for sufficiently long times. The equation for q_1 now reduces to $dq_1/dt = (\lambda_1 + ab/\lambda_2)q_1$ so that $q_1(t) = A\exp((\lambda_1 + ab/\lambda_2)t)$, again in agreement with the asymptotic result.

Although this example has used linear equations it serves to illustrate the power and limitation of the enslaving principle. If there are two distinct time scales (here $1/\lambda_2$ and $1/\lambda_1$ with $\lambda_2 \gg \lambda_1$) then the solution found using the enslaving principle is the same as found from the exact solution for sufficiently large times ($\lambda_2 t \gg 1$). The time has to be sufficiently large to allow the second system (q_2) to relax to its enslaved value bq_1/λ_2.

CHAPTER 6

1. To obtain a dispersion relation we assume a solution for $\phi(t)$ proportional to $\exp(-iwt + ikx)$, substitute into the equation and obtain the condition that the equation is satisfied namely

$$-iw + a + bk^2 = 0 \ .$$

This is the dispersion relation. It is the two-dimensional analogue of an indicial equation. (It is purely conventional to take the argument of the exponential to be $-iwt + ikx$. Any other linear combination of x and t may be used although, of course, the dispersion relation will change. If the equation applies for all x and is bounded then k must be real.)

2. We may write using (6.3) for the case of one mode and the dispersion relation obtained in question 1 that

$$P(x,t) = \frac{1}{2\pi} \int_{-\infty}^{+\infty} dk \ e^{ikx - (a + bk^2)t} \ .$$

Since $-btk^2 + ikx = -bt(k - ix/2bt) - x^2/4bt$ we change the variable in the integrand to 1 where $1 = k - ix/2bt$. Then

$$P(x,t) = \frac{1}{2\pi} e^{-at - x^2/4bt} \int_{-\infty}^{+\infty} dl \ e^{-btl^2}$$

$$= (e^{-at} e^{-x^2/4bt})/2\sqrt{b\pi t} \ .$$

The nth spatial moment M_n is defined by

$$M_n(t) = \int_{-\infty}^{+\infty} x^n \ P(x,t)dx \ .$$

Notice that from the definition of P the normalization condition $M_0 = +1$ is automatically satisfied.

Also note that for n odd the above integral is zero since the integrand is then an odd function of x. For n even, $2m$ say, we make the change of variable in the integrand and write $x^2/4bt = y^2$ in which case

$$M_{2m}(t) = (4^m(bt)^m/\sqrt{\pi})e^{-at} \int_{-\infty}^{+\infty} y^{2m} \ e^{-y^2} \ dy \ .$$

The integral is now just a number depending on m but independent of t, so the full time dependence of M_n is given by the prefactor. This illustrates the usefulness of scaling. The integral can be evaluated and is in fact equal to $\Gamma(m + \frac{1}{2})$ where $\Gamma(x)$ is the gamma function. ($\Gamma(\frac{1}{2}) = \sqrt{\pi}$ and $\Gamma(m + \frac{1}{2}) = (m - \frac{1}{2})\Gamma(m - \frac{1}{2})$ all $m > 0$.)

3. Firstly we assume that $\phi(x, t)$ is a function of $y(= x - Vt)$ only where V is some constant. Then since $\partial^2\phi/\partial t^2$ becomes $V^2 d^2\phi/dy^2$ and $\partial^2\phi/\partial x^2 = d^2\phi/dy^2$ the partial equation reduces to an ordinary one, namely

$$(V^2 - 1)\frac{d^2\phi}{dy^2} + a\phi - b\phi^3 = 0 \ .$$

This may be integrated once to give

$$\frac{(V^2 - 1)}{2}\left(\frac{d\phi}{dy}\right)^2 + \frac{a}{2}\phi^2 - \frac{b\phi^4}{4} = 0 \ .$$

A constant of integration has been put to be zero since the solution we are looking for is soliton in nature so that ϕ and $(d\phi/dy)$ approach zero as $y \to \pm\infty$.

By simply sketching the effective potential $V(\phi) = ((a/2)\phi^2 - b\phi^4/4)/(V^2 - 1)$ (see section 4.4) for the case of a and b both positive it is readily found that bounded solutions for ϕ only exist for $V^2 < 1$. Then we may write

$$\int \frac{d\phi}{\phi\sqrt{\dfrac{a}{2} - b\phi^2/4}} = y\ \sqrt{\frac{2}{1 - V^2}} \ ,$$

where a possible phase factor (constant of integration) has been put to be zero purely for convenience. The integral can be evaluated to give

$$\sqrt{b/2a}\ \cosh^{-1}(\sqrt{2a/b}/\phi) = -y\sqrt{b}/\sqrt{2(1 - V^2)}$$

so that finally

$$\phi(x, t) = \sqrt{2a/b}\ \text{sech}\ ((\sqrt{a/(1 - V^2)})(x - Vt)) \ .$$

4. The analysis proceeds in a manner similar to that used to answer question 3. The corresponding ordinary differential equation takes the form

$$(V^2 - 1) \frac{d^2\phi}{dy^2} = \sin\phi \ .$$

Note that this is of the same form as the equation describing the simple pendulum, namely equation (4.1), but with y replacing t and $(1 - V^2)$ replacing g/l. The equation may be integrated once to give

$$\frac{(V^2 - 1)}{2} \left(\frac{d\phi}{dy}\right)^2 = K - \cos\phi \ ,$$

where K is some constant. Without any loss of generality we may take $V^2 > 1$ as otherwise a change of ϕ to $\phi + \pi$ would convert the equation for $V^2 < 1$ to the one we solve.

To obtain a pulse-like solution we must choose K appropriately, so that the phase plane trajectory passes through the heteroclinic points (see Fig. 4.1). Thus $K = 1$ and we have

$$\int^\phi \frac{d\phi}{\sqrt{1 - \cos\phi}} = y\sqrt{2/(V^2 - 1)} \ .$$

Changing to half angles so that $1 - \cos\phi = 2\sin^2(\phi/2)$ allows the integral to be evaluated to give

$$\ln|\tan(\phi/4)| = y\sqrt{2/(V^2 - 1)} \equiv \alpha y$$

Thus $\tan(\phi/4) = e^{\alpha y}$ so that $\cos(\phi/4) = 1/\sqrt{1 + e^{2\alpha y}}$ and finally

$$\cos(\phi/2) = - \tanh(\alpha y) \ .$$

For $x \to \pm\infty$, $y \to \pm\infty$, $\tanh(\alpha y) \to \pm 1$ and so ϕ changes from $4n\pi$ to $2(2n + 1)\pi$ (n any integer).

The pulse-like nature of this solution is more readily seen by considering $\partial\phi/\partial x = d\phi/dy$. The equation above for ϕ gives

$$\frac{1}{2}\frac{d\phi}{dy} \sin(\phi/2) = \alpha \ \text{sech}^2(\alpha y)$$

and since $\sin(\phi/2) = \text{sech}(\alpha y)$ we have

$$\frac{\partial \phi}{\partial x} = 2\sqrt{2/(V^2 - 1)} \; \text{sech}(x - Vt) \;,$$

a truly pulse-like solution.

5. Comparison of the given equation with (6.15) shows that we may consider (6.15) as appropriate to a solution of the KdV equation at time $t = 0$ of the form $U(x, t = 0) = -2\text{sech}^2 x$, the form required by the question. The value of E is then -1. Since it is the only bounded solution the potential $u(x, t)$ at $t = 0$ is reflectionless and $R(k, t = 0) = 0$. This implies that $R(k, t) = 0$ for all time. With $E = -1$ we have one value of k, namely unity, so from the analysis following equation (6.18) we may write

$$\Psi_a(x, t) = \overline{C} \; e^{4t - x} \;,$$

so in particular $\Psi_a(x, o) = \overline{C} \; e^{-x}$. However, for $t = 0$ we have an explicit solution $(1/\sqrt{2}) \; \text{sech} x$, which in the asymptotic region $x \to +\infty$ goes to $\sqrt{2} e^{-x}$. Thus $\overline{C} = \sqrt{2}$, which gives for the quantity $B(y, t)$ which appears in equation (6.20) the value $2\exp(8t - y)$. With this form it is natural to assume that the solution of (6.20) is of the form $K(x, y, t) = A(x, t)\exp(-\alpha y)$ where α is a constant to be determined. Substitution of this form into (6.20) shows for consistency that $\alpha = 1$ and

$$A(x, t) = -2e^{8t - x}/(1 + e^{8t - 2x}) \;.$$

Finally using this expression in (6.19) gives

$$u(x, t) = -2\frac{d}{dx} \left\{ \frac{-2 \; e^{8t - 2x}}{1 + e^{8t - 2x}} \right\}$$

$$= 4\frac{d}{dx} \; (1/(1 + e^{2x - 8t}))$$

$$= -2\text{sech}^2(x - 4t) \;.$$

This is of the form of the one-soliton solution as given (6.10) with $V = 4$.

6. Assuming a solution of the equation which is a function of $y(= x - Vt)$ only the partial differential equation reduces to

$$V\frac{du}{dy} + 6u\frac{du}{dy} - (\alpha/c) \; V\frac{d^3u}{dy^3} = 0 \;.$$

This may be integrated using the method used in section (6.4) to solve the KdV equation. Alternatively note that by writing $y = \sqrt{\alpha V/c}\, Z$ the above reduces to

$$- (V + 6u) \frac{du}{dZ} + \frac{d^3u}{dZ^3} = 0 \ ,$$

which is the same as the equation for the KdV equation except that Z replaces y. Thus the solution of the above equation is simply as given by (6.10) but with $x - Vt\, (\equiv y)$ being replaced by $Z = \sqrt{c/\alpha V}\, (x - Vt)$. Thus we have the pulse-like solution

$$u(x,t) = - (V/2)\mathrm{sech}^2 \left(\sqrt{\frac{c/\alpha}{2}}\, (x - Vt) \right) \ .$$

Although this is of the same general form as the solution of the KdV equation (6.10) it differs in one important aspect. Here the width of the pulse is independent of the velocity, unlike the KdV solution where the width is inversely proportional to V. This different scaling should be borne in mind when modelling a real physical situation.

7. Substitution of the assumed form of solution into the non-linear Schrödinger equation (1.6) gives

$$\frac{d^2\phi}{dx^2} - w\phi + \beta\phi^3 = 0 \ .$$

One integration gives

$$\left(\frac{d\phi}{dx} \right)^2 - w\phi^2 + \beta\phi^4/2 = 0 \ ,$$

where the integration constant is zero as both $(d\phi/dx)$ and ϕ go to zero together, as $x \to \pm \infty$. This is of the form of the equation integrated in the answer to question 3 ($w = a/(1 - V^2)$ and $\beta = b/(1 - V^2)$) and using the results obtained there we may write the solution of the NLS in the form

$$\Psi(x,t) = \sqrt{2w/\beta}\ e^{iwt}\, \mathrm{sech}\, (\sqrt{w}\, x) \ .$$

8. Substitution of the assumed form of solution into the difference equation readily leads to the equation

$$-iw + a - (b/\Delta^2)(e^{ik\Delta} - 2 + e^{-ik\Delta}) = 0 \ ,$$

as the condition that a non-trivial solution exists. This may be written in the form

$$-iw + a + bk^2 \left[\frac{\sin((k\Delta/2)^2)}{(k\Delta/2)^2} \right] = 0 \ .$$

Comparison with the dispersion relation obtained for the continuous case (question 1) shows that the difference is simply the bracketed term in the above form. Since for $x \rightarrow 0$, $\sin x/x \rightarrow 1$ we see that the dispersion relations are identical in the limit $k\Delta/2 \rightarrow 0$.

An important difference between the two cases is that in the discrete case w remains bounded for all k so that large wave numbers are much less damped.

CHAPTER 7

1. We assume a solution of the form $\phi_n = A\mu^n \equiv A\exp(nk)$ where $k = ln\mu$. Substitution into the equation gives the condition

$$a\mu + b + c/\mu = 0 \ ,$$

which may be solved to give

$$\mu_\pm = (-b \pm \sqrt{b^2 - 4ac})/2a \ .$$

Then by analogy with the treatment of linear differential equations we can write the complete solution of the equation in the form

$$\phi_n = A_+ e^{nk_+} + A_- e^{nk_-} \ ,$$

where $A_\pm$ are arbitrary constants.

2. Imposition of the boundary condition that $\phi_0 = 0$ gives $A_+ = -A_- = A$ so that one can write

$$\phi_n = 2A \ e^{-bn/2a} \sin(n\sqrt{4ac - b^2}/2a) \ .$$

The other boundary condition $\phi_N = 0$ imposes the condition

$$\sin(N\sqrt{4ac - b^2}/2a) = 0 \ ,$$

or

$$(4ac - b^2) = (2ma\pi/N)^2 \ ,$$

where m is an integer. This is an eigenvalue equation, namely a relation between the constants a, b and c which must be satisfied as otherwise there is no non-trivial solution to the difference equation taken with these boundary conditions.

It is instructive to compare this solution with that for the differential equation

$$\alpha \frac{d^2\phi}{dx^2} + \beta \frac{d\phi}{dx} + \gamma\phi = 0 \ ,$$

with $\phi = 0$ at $x = 0$ and $x = L$. The general solution is

$$\phi = B_+ e^{\lambda_+ x} + B_- e^{\lambda_- x} \ ,$$

where

$$\lambda_\pm = (-\beta \pm \sqrt{\beta^2 - 4\alpha\gamma})/2\alpha \ .$$

Imposition of the boundary conditions gives $B_+ = -B_- = B$ and the eigenvalue equation

$$(4\alpha\gamma - \beta^2) = (2\alpha m\pi/L)^2 \ .$$

This illustrates the close analogy which exists between linear difference and linear differential equations. This analogy is lost when one considers non-linear equations.

3. A simple fixed point of a difference equation is defined by demanding that $x_{n+1} = x_n = \bar{x}$ say for all n. In the present case we then have

$$\bar{x} = a\sin(\pi \bar{x}) \ .$$

One solution is always $\bar{x} = 0$. It is not possible to write down a simple analytic expression for the other solutions. However, the nature of these solutions can be easily found by sketching both the left- and right-hand side of the above equation as a function of $\bar{x}$. Solutions to the equation exist when these two sketches intersect. They always intersect at $\bar{x} = 0$ and if $a\pi > 1$ there are two other intersections symmetrically placed about the origin. To consider the stability of these fixed points we write $x_n = \overline{x} + \delta x_n$, substitute into the basic equation, expand in δx_n and neglect all terms of $O(\delta x_n^2)$. That is we linearize to give

$$\delta x_{n+1} = a\pi \cos(\pi\overline{x}) \, \delta x_n \ .$$

Then for a fixed point to be stable we must have $|a\pi \cos(\pi\overline{x})| < 1$. Thus the fixed point loses its stability at $a\pi = 1$, the condition for the appearance of the other two fixed points.

This transition can be studied analytically by noting that for $a\pi \simeq 1$ all three fixed points lie near $\overline{x} = 0$. In this case the $\sin(\pi\overline{x})$ term in the equation for the fixed points may be expanded in powers of $\overline{x}$ to give

$$\overline{x} \simeq a\pi\overline{x} \left(1 - \frac{(\pi\overline{x})^2}{6} \right) \ .$$

One solution is $\overline{x} = 0$ and the other two are

$$(\pi\overline{x})^2 \simeq 6 \left(1 - \frac{1}{a\pi} \right) \ ,$$

obviously only existing for $a\pi > 1$. The stability condition can also be expanded and since $\cos(\pi x) \simeq 1 - \frac{1}{2}(\pi x)^2$ the condition reduces to $|3 - 2a\pi| < 1$. Thus as $a\pi$ increases through unity the $\overline{x} = 0$ state loses its stability and a bifurcation to the other two stable states occurs. This is parallel to the behaviour of the logistic map equation (7.1) for λ near 3.

The above analysis also shows that the two simple fixed points lose their stability for $a\pi > 2$, heralding the bifurcation sequence as found for the logistic map. The criterion $a\pi = 2$ is only an approximation as it was obtained from an expansion procedure which was good for $a\pi \simeq 1$.

Further study of the bifurcation sequence which eventually leads to chaos can only be done numerically and is left as an interesting computational exercise.

4. The fixed points are defined by $\overline{x} = F(\overline{x}, \lambda)$ and it is seen that $\overline{x} = 0$ is always a solution. Linear stability analysis about this fixed point gives $\delta x_{n+1} = F'(\overline{x}, \lambda)\delta x_n = \lambda\delta x_n$ so that $\overline{x} = 0$ is a stable fixed point for $\lambda < 1$. Simple graphical analysis shows that for $\lambda < 1$ this is the only fixed point. For $\lambda > 1$ a fixed point exists such that $\overline{x} > \frac{1}{2}$ so must satisfy $\overline{x} = \lambda(1 - \overline{x})$, that is $\overline{x} = \lambda/(1 + \lambda)$. Linear stability analysis leads to $\delta x_{n+1} = -\lambda\delta x_n$ so that this fixed point which only exists for $\lambda > 1$ is always unstable. Thus for $\lambda < 1$ there is one stable fixed point at the origin but for $\lambda > 1$ the simple fixed points are unstable.

A similar analysis can be carried out for period-two fixed points. These are defined as $\overline{x}$, $\overline{y}$ where $\overline{y} = \lambda\overline{x}$ and $\overline{x} = \lambda(1 - \overline{y})$ with $\overline{x} < \frac{1}{2}$ and $\overline{y} > \frac{1}{2}$. These can be solved to $\overline{x} = \lambda(1 + \lambda^2)$, $\overline{y} = \lambda^2/(1 + \lambda^2)$ and have solutions satisfying the restrictions on $\bar{x}$ and $\bar{y}$ for $\lambda > 1$. To carry out the stability analysis we write $x_n = \bar{x} + \delta x_n$, $x_{n+1} = \bar{y} + \delta x_{n+1}$ and $x_{n+2} = \bar{x} + \delta x_{n+2}$. Substitution into the equation $x_{n+1} = F(x_n, \lambda)$ gives $\delta x_{n+1} = \lambda \delta x_n$ and $\delta x_{n+2} = -\lambda \delta x_{n+1}$ so that $\delta x_{n+2} = -\lambda^2 \delta x_n$ and since $\lambda > 1$ the period-two fixed point is unstable.

At the expense of a little algebraic manipulation higher-order fixed points may be studied. These exist for $\lambda > 1$ but are always unstable. The tent map is an example of a map which is always chaotic ($\lambda > 1$). This is verified by evaluating the Liapunov number, which is always positive for $\lambda > 1$.

5. For the tent map as defined in question 4 we have $|F'(x,\lambda)| = \lambda$ for all x so that equation (7.12) for the asymptotic distribution function is

$$P(x) = (P(x_1) + P(x_2))/\lambda \ ,$$

where $x = \lambda x_1 = \lambda(1 - x_2)$. Such equations are difficult to solve but for $\lambda = 2$ it is seen by inspection that $P(x) = 1$ for all x is a solution.

Then from equation (7.14) the Liapunov number is given by

$$\mu = \int_1^1 dx \ \ln 2 = \ln 2 \ .$$

Since $\mu > 0$ the system is chaotic, a result consistent with the analysis of question 4 where no stable fixed points of the map were found to exist.

6. The condition for a map to have a tangent bifurcation is $\bar{x} = F(\bar{x}, \eta)$ and $1 = F_x(\bar{x}, \eta)$ which in the present case reduce to the conditions $\bar{x} = \eta - 1/\bar{x}$ and $1 = 1/\bar{x}^2$. These give $\bar{x} = \pm 1$ and hence $\eta = \pm 2$. This result is easily checked by sketching $F(x, \eta)$ as a function of x and seeing that for $\eta = 2$ the line $y = -x$ is tangent to it at the point $x = 1$ ($\eta > 0$).

7. From the general form of solution given in the question we may write

$$x_{n+1} = \cos a - \sin a \ \tan((n+1)a + b) \ .$$

Using simple trigonometric identities we express the right-hand side of the above equation in terms of x_n. Thus

$$x_{n+1} = \cos a - \sin a \left\{ \frac{\tan(na+b) + \tan a}{1 - \tan a \ \tan(na+b)} \right\}$$

$$= \cos a - \sin a \ \frac{\left\{ \dfrac{(\cos a - x_n)}{\sin a} + \tan a \right\}}{\left\{ 1 - \dfrac{\tan a}{\sin a}(\cos a - x_n) \right\}}$$

$$= \cos a - \left\{ \frac{1 - (\cos a)\, x_n}{x_n} \right\}$$

$$= 2\,\cos a - \frac{1}{x_n}\ .$$

Thus we must make the identification $\eta = 2\cos a$.

For $\eta < 2$ we have that a is real and the solution to the difference equation is as given in the question. For $\eta > 2$ we must take a to be purely imaginary and write $a = i\bar{a}$. Then $\eta = 2\cosh\bar{a}$ and the solution now takes the form

$$x_n = \cosh\bar{a} + \sinh\bar{a}\ \tanh\,(n\bar{a} + \phi)\ ,$$

with ϕ and x_n both real.

8. Taking $x_n = s + by_n$ and substituting from the map gives

$$x_{n+1} = s + by_{n+1} = s + b\lambda y_n/(1 + (\lambda - 1)y_n)$$
$$= s + \lambda(x_n - s)/(1 + (\lambda - 1)(x_n - s)/b)\ .$$

It is essential that the denominator in the above equation is proportional to x_n. Hence we choose $1 - (\lambda - 1)s/b = 0$ in which case the above map reduces to

$$x_{n+1} = s(1 + \lambda) - \lambda s^2/x_n\ ,$$

so we make the identification $\eta = s(1 + \lambda)$ and $\lambda s^2 = 1$. With $\lambda = e^\alpha$ we have the condition that $\eta = 2\cosh(\alpha/2)$.

Comparison with the solution of question 6 shows that since $\eta > 2$ we make the identification $a = \alpha/2$. With $y_n = (x_n - s)/b$ we have, after a little algebraic manipulation, that

$$y_n = (1 + \tanh\,(n\alpha/2 + \phi))/2$$

($s = \exp(-\alpha/2)$ and $b = 2\sinh(\alpha/2) = +2\sinh a$).

Since for large argument $\tanh x \simeq 1 - 2e^{-2x}$ the above form for y_n as $n \to +\infty$ reduces to

$$y_n \simeq 1 - Ae^{-\alpha n}\ ,$$

where $A = \exp(-2\phi)$ is an arbitrary constant.

This result we interpret as saying that the fixed point $y_n = 1$ is stable since any perturbation (that is finite value of A) decays as $n \to \infty$. Of course this agrees with a linear stability analysis of the integral equation about the fixed point $y = 1$.

CHAPTER 8

1. The fixed points are defined by the conditions $p_{n+1} = p_n = p$ and $\theta_{n+1} = \theta_n|$ module 2π. The simplest is $p = \theta = 0$. Linearization of the map (8.1) about this point gives

$$\delta p_{n+1} = \delta p_n - K \delta \theta_n$$
$$\delta \theta_{n+1} = \delta \theta_n + \delta p_{n+1} \ .$$

These equations may be solved by assuming solutions proportional to μ^n. This leads to the consistency condition

$$(\mu - 1)^2 = -K\mu \ .$$

Writing $\mu = \exp(i\eta)$ changes this condition to $\cos\eta = 1 - K/2$. Thus for $K < 4$, η is real and the fixed point is stable; it is a centre. For $K > 4$, η is purely imaginary and the fixed point now becomes a saddle.

Another type of fixed points exist where $p = \pi$ and $\theta_n = n\pi$, n any integer. Linear stability analysis about these fixed points gives

$$(\mu - 1)^2 = -K\mu \cos n\pi \ ,$$

so with $\mu = \exp(\pm i\eta)$ we have $\cos\eta = 1 - \dfrac{K}{2} \cos(n\pi)$.

For n odd $\cos n\pi = -1$ and η is imaginary and the fixed point is a saddle for all values of K ($K > 0$). For n even the situation is identical to that for $n \equiv 0$, a centre for $K < 4$ and a saddle for $K > 4$.

Thus for $K > 4$ there are no stable simple fixed points and this suggests that for $K > 4$ the solution should be chaos. This is borne out by numerical simulations, though there are non-chaotic regions in the phase plane associated with higher-order periodic orbits.

In the continuum approximation the fixed points are $p = 0$ and $\sin\theta = 0$. Linearization of equation (8.7) about these points gives

$$\frac{d^2\delta\theta}{dt^2} + (K/\Delta^2) \cos\theta \ \delta\theta = 0 \ ,$$

so that $\delta\theta \propto \exp(\pm\sqrt{K\cos\theta}\ \tau/\Delta)$. Thus for $\theta = 2n\pi$ we have a saddle point for all K whilst for $\theta = (2n+1)\pi$ we have a centre for *all* K.

2. We have to $O(K^2)$ that $\theta_n = a + pn + K\theta_n^{(1)}$ and $p_n = p + Kp_n^{(1)}$ with $\theta_n^{(1)}$ and $p_n^{(1)}$ given by equation (8.10). To obtain the form of the relationship between p_n and θ_n we use the above expressions to remove the explicit n dependence. The equation for θ_n may be rewritten in the form $a + pn = +\theta_n + O(K)$ and when this is substituted into the expression for p_n we have

$$p_n = p + \frac{K}{2}\ \{\sin\theta_n + \cot(p/2)\cos\theta_n\} + O(K^2)\ .$$

   Now θ_n is a variable that can take a range of values ($\theta_n \approx a + pn$) as n is varied so the iterates of the map fall on the line $p(\theta) \equiv p_n(\theta)$ as obtained above. However, because θ_n is not a continuous variable this line is not filled in a continuous manner. For example if $p = 2\pi/m$ then since θ_n is module 2π it only takes m distinct values and only m distinct points occur in the map, although they fall on the line $p(\theta)$.

   A slight change of p so that it is an irrational multiple of 2π would mean that the iterates of the map would then appear as a continuous line, namely $p(\theta)$. Thus for small K one expects the phase plane to be interlaced with continuous lines (irrational ps) and a series of points lying on a line (rational ps).

   To next order in K these lines are given by equation (8.15) with $a_n = a_{n-1}$ and are shown in Fig. 8.2.

3. This expression for θ may be shown to satisfy equation (8.7) by direct substitution. Alternatively the equation may be integrated once to give

$$\left(\frac{d\theta}{d\tau}\right)^2 = \frac{2K}{\Delta^2}\ (1+\cos\theta)\ ,$$

where the constant of integration has been chosen such that for $\theta = \pm\pi$, $d\theta/dt = 0$. Then we have

$$\int \frac{d\theta}{\sqrt{1+\cos\theta}} = \sqrt{2K/\Delta^2}\ \tau\ .$$

The integral may be evaluated by changing to half angles and one then finds that (see question 4 of Chapter 6)

$$\theta = -\pi + 4\tan^{-1}\ (\exp(\sqrt{K/\Delta^2}\ \tau))\ .$$

Simple differentiation with respect to t ($\tau = \Delta t$) gives

$$\frac{d\theta}{dt} = 2\sqrt{K} \operatorname{sech}(\sqrt{K}\, t) ,$$

so that the maximum value of $d\theta/dt$ is $2\sqrt{K}$ and hence the width of the heteroclinic orbit passing from $(-\pi, 0)$ to $(\pi, 0)$ is $4\sqrt{K}$.

This scaling of $d\theta/dt$ with $\sqrt{K}$ can be seen directly by inspection of the above equation involving $d\theta/dt$.

4. The simple fixed points of the Hénon map satisfy the equations

$$\bar{x} = 1 - a\bar{x}^2 + b\bar{y}$$
$$\bar{y} = \bar{x} .$$

These may be solved to give two distinct fixed points

$$\bar{x}_\pm = ((b-1) \pm \overline{\sqrt{(1-b)^2 + 4a}})/2a .$$

Linear stability analysis gives on writing $x_n = \bar{x} + \delta x_n$ and $y_n = \bar{x} + \delta y_n$ that

$$\delta x_{n+1} = -2a\,\bar{x}\,\delta x_n + b\delta y_n$$
$$\delta y_{n+1} = \delta x_n .$$

These equations have solutions with δx_n and δy_n proportional to μ^n where

$$\mu(\mu + 2a\bar{x}) = b$$
$$\mu = -a\bar{x} \pm \sqrt{(a\bar{x})^2 + b} .$$

For small $b(b/(a\bar{x})^2 \ll 1)$ the square root may be expanded to give to lowest order $\mu = b/2a\bar{x}$ or $\mu = -2a\bar{x}$. Also for small b we may write $\bar{x} = (-1 \pm \sqrt{1+4a})/2a$ so the values of μ are

$$\mu = \frac{b}{-1 \pm \sqrt{1+4a}} \quad \text{and} \quad \mu = 1 \pm \sqrt{1+4a} .$$

The condition for stability is $|\mu| < 1$, which is always satisfied for the value of μ proportional to b. The fixed point $\bar{x} = (-1 + \sqrt{1+4a})/2a$ is stable for

$\sqrt{1 + 4a} - 1 < 1$, that is $a < 3/4$, whilst the other fixed point is always unstable; that is, for $a > 3/4$ both fixed points are unstable.

5. The period-two fixed points of the Hénon map labelled by (x_0, y_0) and (x_1, y_1) satisfy the equations $x_1 = 1 - ax_0^2 + by_0$, $y_1 = x_0$ and $x_0 = 1 - ax_1^2 + by_1$, $y_0 = x_1$. Eliminating x_1 and y_1 gives $\gamma y_0 = 1 - ax_0^2$ and $\gamma x_0 = 1 - ay_0^2$ where $\gamma = 1 - b$. These may be reduced to a quartic equation in x_0 namely

$$a^3 x_0^4 - 2a^2 x_0^2 + \gamma^3 x_0 + a - \gamma^2 = 0 \ .$$

To solve this equation we note that the equation must contain as solution the two fixed points associated with period-one, that is where $\gamma x_0 = 1 - a x_0^2$ (see question 4). Thus we factor the above quartic into the quadratic $ax_0^2 + \gamma x_0 - 1$ and another quadratic. This then leads to the equation

$$a^2 x_0^2 - a\gamma x_0 + \gamma^2 - a = 0 \ ,$$

for the period-two fixed points. Real solutions to this equation exist if $a > 3\gamma^2/4$.

Note that for $b \ll 1$ this condition is simply $a > \frac{3}{4}$, which is the condition obtained in the answer to question 4 for a simple fixed point to go unstable. This is another example of a bifurcation.

APPENDICES

1. Consider the integral

$$I = \int_{q_1}^{q_2} P_m^+ L P_n \, dq \ ,$$

which from (A1.2) takes the form

$$I = \int_{q_1}^{q_2} P_m^+ \left\{ D \frac{d^2 P_n}{dq^2} - \frac{d}{dq}(HP_n) + \lambda_n P_n \right\} dq \ .$$

Integration by parts of the first two terms and using the condition that both P_m^+ and P_n are zero for $q = q_1, q_2$ gives

$$I = \int_{q_1}^{q_2} P_m \left\{ D \frac{d^2 P_m^+}{dq^2} + H \frac{dP_m^+}{dq} + \lambda_n P_m^+ \right\} dq \ .$$

But from the definition of the adjoint operator

$$I = \int_{q_1}^{q_2} P_m L^+ P_n \, dq$$

and indentifying the above two expressions for I leads to the equation for the adjoint operator namely (A1.3).

It is important to note that the definition of the adjoint operator is only complete when appropriate boundary conditions are specified.

2. To apply the perturbation theory to the problem we write

$$x(t) = x_0 + \varepsilon x_1 + \varepsilon^2 x_2 + O(\varepsilon^3) \, ,$$

and consider the xs to be functions of the many times $\tau_0 = t$, $\tau_1 = \varepsilon t$. Note that we take $\tau_1 = \varepsilon t$ and not $\varepsilon^2 t$ as in the example in Appendix A2. Now we have

$$\frac{d^2 x}{dt^2} = \frac{\partial^2 x_0}{\partial \tau_0^2} + \varepsilon \left(\frac{\partial^2 x_1}{\partial \tau^2} + 2 \frac{\partial^2 x_0}{\partial \tau_0 \partial \tau_1} \right) + O(\varepsilon^2) \, .$$

Substitution of these expansions into the equation under discussion and equating terms independent of ε and those proportional to ε separately to zero gives

$$\frac{\partial^2 x_0}{\partial \tau_0^2} + w^2 x_0 = 0$$

and

$$\frac{\partial^2 x_1}{\partial \tau_0^2} + w^2 x_1 = x_0 - 2 \frac{\partial^2 x_0}{\partial \tau_0 \partial \tau_1} \, .$$

The solution of the first equation is

$$x_0 = A \cos (w \tau_0 + \phi) \, ,$$

where A and ϕ are independent of τ_0 but may be functions of τ_1. With this result the second equation becomes ($\cos^3 x = (3\cos x + \cos 3x)/4$)

$$\frac{\partial^2 x_1}{\partial \tau_0^2} + w^2 x_1 = 2w \left\{ \frac{\partial A}{\partial \tau_1} \sin (w\tau_0 + \phi) + A \frac{\partial \phi}{\partial \tau_1} \cos (w\tau_0 + \phi) \right\}$$

$$+ (A^3/4) \left\{ \cos(3(w\tau_0 + \phi)) + 3\cos(w\tau_0 + \phi) \right\} .$$

The resonant or secular terms which give rise to non-periodic behaviour of x_1 are those proportional to $\sin(w\tau_0 + \phi)$ or $\cos(w\tau_0 + \phi)$. Thus if we demand that x and hence x_1 is periodic in time these terms must be put to zero. Of course without the multiple time terms this would not be possible. This condition gives $\partial A/\partial \tau_1 = 0$ and

$$2w \frac{\partial \phi}{\partial \tau} = -3A^2/4 .$$

The equation for x_1 without the secular terms is now readily solved and the above results combined to give the final result, after putting $\tau_1 = \varepsilon t$, that

$$x(t) = A\cos(\overline{w}t + \overline{\phi}) - \varepsilon(A^3/32w^2)\cos(3(\overline{w}t + \overline{\phi})) + O(\varepsilon^2) ,$$

where $\overline{w} = w - \varepsilon 3A^2/8w$ and A and $\overline{\phi}$ are constants to be determined by initial conditions.

In this case it is seen that the non-linearity has introduced a frequency shift with the shift depending on the amplitude. Such shifts are a common feature of non-linearity.

3. Here we proceed as in question 2. The equation for x_0 is the same, but that for x_1 takes the form

$$\frac{\partial^2 x_1}{\partial \tau_0^2} + w^2 x_1 = 2w \left\{ \frac{\partial A}{\partial \tau_1} \sin(w\tau_0 + \phi) \right.$$

$$\left. + A \frac{\partial \phi}{\partial \tau_1} \cos (w\tau_0 + \phi) \right\}$$

$$+ (A^2/2) \left\{ 1 + \cos(2(w\tau_0 + \phi)) \right\} .$$

Now we note that to this order the non-linearity does not introduce any resonant terms so that $\partial A/\partial\tau_1 = 0$ and $\partial\phi/\partial\tau_1 = 0$. The equation for x_1 may be solved to give

$$x_1 = (A^2/2w^2)\{1 - \tfrac{1}{3}\cos(2(w\tau_0 + \phi))\}\ ,$$

which is periodic. To proceed to next order it is necessary to introduce a further time variable $\tau_2 = \varepsilon^2 t$. (Note that with hindsight one could have introduced a time variable $\tau_1 = \varepsilon^2 t$ from the start. This was the reason for the ordering chosen to discuss the problem in Appendix 2.)

Proceeding to next order gives

$$\frac{\partial^2 x_2}{\partial\tau_0^2} + w^2 x_2 = 2w\left\{\frac{\partial A}{\partial\tau_2}\sin(w\tau_0 + \phi)\right.$$

$$\left. + A\frac{\partial\phi}{\partial\tau_2}\cos(w\tau_0 + \phi)\right\}$$

$$+ (A^3/6w^2)\ \{5\cos(w\tau_0 + \phi)$$

$$- \cos(3(w\tau_0 + \phi))\}\ ,$$

since the non-linear term on the right-hand side is simply $2x_0x_1$ to this order. The removal of secular terms demands that $\partial A/\partial\tau_2 = 0$ and $\partial\phi/\partial\tau_2 = -5A^2/12w^2$. After the removal of secular terms x_2 may be solved and found proportional to $\cos(3(w\tau_0 + \phi))$. All this leads to the final solution

$$x(t) = A\cos(\overline{w}t + \overline{\phi}) + \varepsilon(A^2/2w^2)(1 - \frac{1}{3}\cos(2(\overline{w}t + \overline{\phi})))\ ,$$

$$+ \varepsilon^2\ (A^3/96w^2)\cos(3(\overline{w}t + \overline{\phi})) + O(\varepsilon^3)\ ,$$

where $\overline{w} = w - 5A^2/12w^2$ and A and $\overline{\phi}$ constants at least to this order in ε.

Both the examples considered in this question and the previous one have shown that the non-linearity introduces a frequency (or phase) change. An equation where the method gives an equation for the variation of the amplitude is the Van der Pol equation discussed in section (4.3).

4. These equations are in fact the Boussinesq equations describing the spatial and temporary variation of the height h of the surface of a shallow fluid in the gravitational field (g) which has a velocity u. The constant α is proportional to the unperturbed height, h_0, of the fluid.

An ordering scheme must be devised such that the dispersive effects associated with the term $\alpha \partial^3 h / \partial x \partial t^2$ are of the same order as the non-linear effects associated with the terms $u \partial u / \partial x$ and $\partial(hu)/\partial x$. Further all these terms must be small and not enter into the lowest-order theory. Thus we write $h(x,t) = h_0 + \varepsilon^2 h_1 + \varepsilon^4 h_2 + O(\varepsilon^4)$ and $u(x,t) = \varepsilon^2 u_1 + \varepsilon^4 u_2 + O(\varepsilon^6)$ with u_1, u_2, h_1 and h_2 functions of $x_1 = \varepsilon x$, $\tau_1 = \varepsilon t$ and $\tau_2 = \varepsilon^3 t$.

Substituting this expansion into the Boussinesq equations and equating the coefficients of ε^n separately to zero gives to $0(\varepsilon^3)$

$$\frac{\partial h_1}{\partial \tau_1} + h_0 \frac{\partial u_1}{\partial x_1} = 0$$

and

$$\frac{\partial u_1}{\partial \tau_1} + g \frac{\partial h_1}{\partial x_1} = 0$$

and to $O(\varepsilon^5)$

$$\frac{\partial h_2}{\partial \tau_1} + h_0 \frac{\partial u_2}{\partial x_1} = -\frac{\partial h_1}{\partial \tau_2} - \frac{\partial}{\partial x_1}(h_1 u_1)$$

$$\frac{\partial u_2}{\partial \tau_1} + g \frac{\partial h_2}{\partial x_1} = -\alpha \frac{\partial h_1}{\partial x_1 \partial \tau_1^2} - u_1 \frac{\partial u_1}{\partial x_1} - \frac{\partial u_1}{\partial \tau_2} \ .$$

The equations to $O(\varepsilon^3)$ may be combined to give a wave equation

$$\frac{\partial^2 h_1}{\partial t_1^2} - h_0 g \frac{\partial^2 h_1}{\partial x_1^2} = 0 \ ,$$

which has a solution of the form $h_1(x_1, \tau_1) = p(x_1 - \sqrt{h_0 g}\ \tau_1) + q(x_1 + \sqrt{h_0 g}\ \tau_1)$ where p and q are arbitrary functions. We restrict attention to waves propagating in the positive space direction so that $q \equiv 0$. In this case $h(x_1, \tau_1) = p(\xi)$ and $u_1(x_1, \tau_1) = \sqrt{g/h_0}\ p(\xi)$ where $\xi = x_1 - \sqrt{h_0 g}\ \tau_1$.

The equations to $O(\varepsilon^5)$ are manipulated to remove u_2. The first is differentiated with respect to τ_1, the second with respect to x_1 and then simple algebraic manipulation gives

$$\frac{\partial^2 h_2}{\partial \tau_1^2} - h_0 g \frac{\partial^2 h_2}{\partial x_1^2} = \frac{\partial}{\partial \xi} \left\{ 2\sqrt{h_0 g}\ \frac{\partial p}{\partial \tau_2} + 3g\, p\, \frac{\partial p}{\partial \xi} + \alpha\, h_0^2 g\, \frac{\partial^3 p}{\partial \xi^3} \right\} \ .$$

The left-hand side of this equation can be written in the form $\partial^2\phi/\partial\xi\partial\eta$ where $\eta = x_1 + \sqrt{h_0 g}\ \tau_1$. Then since the right-hand side is a function of ξ only, integration would in general give h_2 to be proportional to η. This would of course mean that h_2 would become unbounded as $\eta \to \infty$, even though $p(\xi)$ was a bounded function of ξ. To remove the difficulty we simply insist that the right-hand side is identically zero and in particular that

$$2\sqrt{h_0 g}\ \frac{\partial p}{\partial \tau_2} + 3gp\ \frac{\partial p}{\partial \xi} + \alpha\ h_0^2\ g\ \frac{\partial^3 p}{\partial \xi^3} = 0 \ .$$

This of course is of the form of the KdV equation and can easily be transformed into the standard form of equation (6.8) by identifying p with $-4\alpha h_0^2 u$, ξ with x and $\alpha h_0^2 \sqrt{g/h_0}\ \tau_2$ with t.

It should be noted that although in the above we treated α as a constant it could have been taken to be any function of the height h since the ordering scheme chosen only requires the value of α for $h = h_0$.

References

Haken, H., *Synergetics*, Springer, Berlin and New York (1978).

Infeld, E. and Rowlands, G., *Non-linear waves, solitons and chaos*, Cambridge University Press (1990).

Mandelbrot, B. B., *The fractal geometry of nature*, Freeman, San Francisco (1982).

Minorsky, N., *Non-linear oscillations*, Van Nostrand, Princeton (1962).

Nayfeh, A. H., *Perturbation methods*, John Wiley, New York (1973).

Poston, T. and Stewart, I. *Catastrophe theory and its applications*, Pitman, London (1978).

For further reading on stochastic equations see

Gardiner, C. W., *Handbook of stochastic methods*, Springer-Verlag, Berlin and New York (1985).

The subject of chaos and in particular one-dimensional maps is treated in a more rigorous mathematical manner by

Schuster, H. G., *Deterministic chaos, an introduction*, Physik Verlag, Weinheim (1984).

The book by

Lichtenberg, A. and Lieberman, M. A., *Regular and stochastic motion*, Springer, Berlin and New York (1983)

is highly recommended for further study of two-dimensional maps. (Attention is paid to how such maps arise in the study of real physical situations.)

The computing package which illustrates much of the material in Chapters 7 and 8 of the present book is by Sprott, J. C. and Rowlands, G. entitled *Chaos demonstrations* (1990) and marketed by the American Institute of Physics under the heading Physics Academic Software.

Index

Index